AutoCAD® 2015
3D Modeling

Publisher: David Pallai
MERCURY LEARNING AND INFORMATION
22841 Quicksilver Drive
Dulles, VA 20166
info@merclearning.com
www.merclearning.com
(800) 232-0223

This book is printed on acid-free paper.

Munir M. Hamad, AUTOCAD® 2015 3D Modeling.
ISBN: 978-1-937585-37-2

Library of Congress Control Number: 2014950128
151617 321

Our titles are available for adoption, license, or bulk purchase by associations, universities, corporations, etc. Digital versions of this title may be purchased at www.authorcloudware. com or other e-vendors. For additional information or companion e-files, please contact the Customer Service Dept. at 1-(800) 232-0223 (toll free) or info@merclearning.com.

PREFACE

Since its inception, AutoCAD has enjoyed a very wide user base and has been the most widely used CAD software since the 1980s. This popularity is due to its logic and simplicity, which makes it very easy to learn. AutoCAD has evolved throughout the years and has become comprehensive software, addressing all aspects of engineering drafting and designing.

This book addresses the 3D environment of AutoCAD and teaches the reader how to create and edit 3D objects in AutoCAD and produce real-world images out of them. This book is ideal for advanced AutoCAD users and college students who have already studied the basics of AutoCAD 2015.

This book is not a replacement for the manual(s) that comes with the software but is considered complementary with its practices and projects, meant to strengthen the knowledge gained and solidify the techniques discussed. Solving all practices and projects is essential because AutoCAD is practical and not theoretical. The book includes 50 "mini-workshops," that complete small projects from concept through actual plotting. Solving all of the workshops will simulate the creation of full projects (architectural and mechanical). Solving these complete projects will help the reader to master the knowledge needed to land a job in today's market. *These projects are presented in both metric and imperial units.*

At the end of each chapter the reader will find "Chapter Review Questions." These are the same sort of questions you might see in an Autodesk exam. The answers to the odd questions are included at the end of each chapter to check your work.

ABOUT THE CD-ROM

The companion disc included with this book contains:

- A link to AutoCAD 2015 Trial version, which is valid for 30 days starting from the day of installation. This version can be used to solve all exercises.

- Practice files, which are essential to working through the projects in the book.

- A folder named "Practices and Projects," which you should copy to the hard drive of your computer.

- Two folders of projects: "Metric" for metric unit projects, and "Imperial" for imperial unit projects.

PREREQUISITES

The author assumes the reader has a thorough understanding of computers and the Windows operating system. You should know how to:

- Create a new file

- Open an existing file

- Save and Save as (files)

- Closing and exiting the software

These commands are nearly the same in all software offerings. The author does not explain these commands other than to illustrate some specific actions in AutoCAD.

AutoCAD 2015 features a dark gray background. All screen captures displayed in this book, however, have been changed to white to improve the presentation.

CHAPTER 1

*Auto*CAD 3D BASICS

In This Chapter

- Using the 3D interface
- Viewing the 3D model
- Using UCS commands

1.1 RECOGNIZING THE 3D ENVIRONMENT

AutoCAD is used to draw 2D objects and 3D models. Working with 3D models is accomplished by using two rules:

- The 3D template files **acad3D.dwt** and **acadiso3D.dwt** are used to set up the 3D environment needed to build a 3D model.

- The **3D Basic** workspace and **3D Modeling** workspace are used to build 3D models. The 3D Modeling workspace is used in all examples in this book, as all AutoCAD 3D commands are included in its tabs and panels.

1.2.2 Using the ViewCube

AutoCAD displays the ViewCube in the upper right corner of each drawing. See the following illustration of the ViewCube:

Clicking a face of the cube provides one of the 2D orthographic preset views. Clicking an upper corner of the cube provides an isometric view similar to SW, SE, NE, and NW. Clicking a lower corner of the cube provides a similar isometric view, but from below. In addition to these views, ViewCube provides a view of the model from each edge of the cube. In total there are twenty six views.

The ViewCube can also be rotated CCW or CW using the compass.

AutoCAD saves one of the views as the Home view, which is displayed when the Home icon is clicked.

Views are oriented to the WCS (World Coordinate System) or the UCS (User Coordinate System). It is important that you are aware of the coordinate system that is in use (this is covered in greater detail at end of this chapter). Click the pop-up list at the bottom of the ViewCube to display view options, such as:

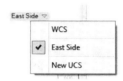

A small triangle in the lower right corner of the ViewCube launches a settings menu. See the following illustration:

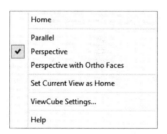

You can perform the following:

- Retain the Home view
- Switch between Parallel (60/30 isometric) view, Perspective view, or Perspective with Ortho Faces view
- Change the Home view to the current view
- Modify ViewCube Settings
- Display ViewCube Help topics

Selecting **ViewCube Settings** displays the following dialog box:

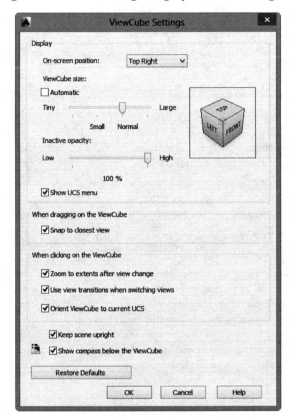

Set the following:

- Location of the ViewCube (default is Top Right)
- ViewCube size (default is Automatic)
- Inactive Opacity (default is 50%)
- Show or hide the UCS menu

The rest of the settings are self-explanatory.

PRACTICE 1-1

AutoCAD 3D Basics

1. Start AutoCAD 2015.

2. Open Practice 1-1.dwg.

3. Using the preset views of ViewCube (try all options), view the 3D model from every angle.

4. Save and close the file.

1.3 ORBITING IN AutoCAD

Orbiting is the dynamic rotation of your model using a mouse. There are three orbiting options in AutoCAD:

- Orbit
- Free Orbit
- Continuous Orbit

To find the three Orbit commands, from the **View** tab, locate the **Navigate** panel, and click the **Orbit** button, as displayed here:

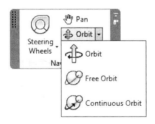

1.3.1 Orbit Command

Selecting the Orbit command changes the mouse cursor to this shape:

The Orbit command enables the rotation of your model around all axes. Views are limited to top and bottom. Press [Esc] to end the command.

You can accomplish the same Orbit rotation by holding [Shift] and clicking and holding the mouse wheel.

1.3.2 Free Orbit

As displayed in the following illustration, an arcball appears, displaying a circle at each quadrant. Zoom and pan using your mouse wheel to center your model within the arcball:

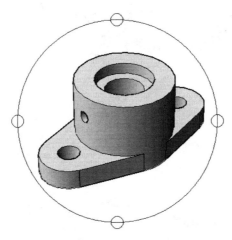

Depending on where your cursor is positioned on the arcball, you will see a different cursor shape.

If your cursor is positioned by a circle on the right or left side of the arcball, ⟳ is displayed as the cursor shape. Click and drag the mouse to the right or left to rotate the shape around the vertical axis.

If your cursor is positioned by a circle on the top or bottom of the arcball, ⟳ is displayed as the cursor shape. Click and drag the mouse up or down to rotate the shape around the horizontal axis.

If your cursor is positioned within the arcball, ⟳ is displayed as the cursor shape, with the same functionality as that described for the Orbit command.

If your cursor is positioned outside the arcball, ↺ is displayed as the cursor shape. Click and drag the mouse to rotate the model around an axis that runs through the center of the arcball and extends beyond the screen. Press [Esc] to end this command.

Another way to access this command is by holding [Shift]+[Ctrl]+Mouse wheel.

1.3.3 Continuous Orbit

Selecting the Continuous Orbit command changes the mouse cursor to this shape:

Using the Continuous Orbit command initiates a continuous rotation of the model. Click and drag your mouse on the model and it will rotate in the direction specified. Press [Esc] to end this command.

1.4 STEERING WHEEL

The Steering Wheel allows you to navigate the model using methods such as zooming, orbiting, panning, walking, etc.

To initiate the Steering Wheel command, from the **View** tab, locate the **Navigate** panel and click the **Steering Wheels** button, as displayed in the following:

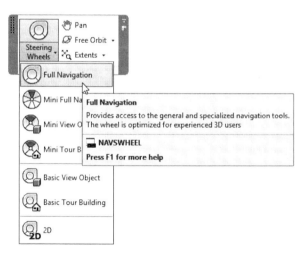

Choose the **Full Navigation** option. The following is displayed:

There are eight viewing commands: four in the outer circle and four in the inner circle.

The following commands are presented in the outer circle:

- Zoom
- Orbit
- Pan
- Rewind

The following commands are presented in the inner circle:

- Center
- Look
- Up/Down
- Walk

1.4.1 Zoom Command

The Zoom command allows you to zoom into the 3D model (using the mouse wheel accomplishes the same). Move your cursor to the desired location of your model, start the Zoom option, click and hold the mouse button, and move your mouse forward and backward to zoom in and out.

1.4.2 Orbit Command

The Orbit command on the Steering Wheel is similar to the Orbit command selected on the Navigate panel. Move your cursor to the desired location, select the Orbit option, click and hold the mouse button, and move the mouse right, left, up, and down to orbit your model.

1.4.3 Pan Command

The Pan command enables panning in 3D. Move your cursor to the desired location, select the Pan option, click and hold the mouse button, and move the mouse right, left, up, and down to pan your model.

1.4.4 Rewind Command

The Rewind command displays a record of all the actions you have taken using the other seven commands. When started, it shows a series of screen shots of the different actions you have taken. Click and drag your mouse to the left to view all of your actions.

1.4.5 Center Command

The Center command allows you to specify a new center point for the screen. Select the Center command, click and hold the mouse button, locate a new center point for the current view, and release the mouse button. The entire display adjusts to represent the new center point.

1.4.6 Look Command

The Look command provides the ability to look around the model from a stationary position. Move the cursor to the desired position, activate the command, click and hold the mouse button, and move the mouse right, left, up, and down to look around the model.

1.4.7 Up/Down Command

The Up/Down command provides a view of your model from above or below. Move the cursor to the desired location and select the Up/Down command. On the vertical scale that appears, click and hold the mouse button, and move the mouse up and down.

1.4.8 Walk Command

The walk command simulates the ability to walk around the model. There are eight directions of motion. The best results are achieved when this is combined with the Up/Down command. Move your cursor to the desired location, select the Walk command, click and hold the mouse button, and move the mouse in one of the eight directions.

A miniature version of the Steering Wheel can be accessed from the **Navigate** panel in the **View** tab. It features the following:

- Mini Full Navigation Wheel displays a small circle with all eight commands.

- Mini View Object Wheel displays a small circle featuring the four commands of the full wheel's outer circle.

- Mini Tour Building Wheel displays a circle featuring the four commands of the full wheel's inner circle.

- Basic View Object Wheel

- Basic Tour Building Wheel

1.5 VISUAL STYLES

Visual styles determine how your objects are displayed on the screen. The default visual style in 3D templates is **Realistic**. Clicking the Visual Style Controls tab in the top left of the screen presents the following menu:

You can also access this menu from the **Home** tab, then the **View** panel, and by clicking the drop-down list:

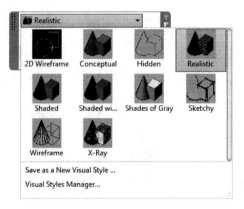

AutoCAD has ten predefined visual styles (more can be made). The predefined styles are:

- 2D Wireframe visual style displays the edges, lineweight, and linetype of objects.

- Wireframe visual style displays the edges of all objects without lineweight or linetype.

- Hidden visual style is similar to Wireframe, but it does not display lines that are obscured from the current viewpoint.

- Conceptual visual style displays objects with smooth shading.

- Realistic visual style displays objects with smooth shading, and with texture if materials are assigned.

- Shaded visual style is similar to Realistic, but without texture.

- Shaded with Edges visual style is similar to Shaded, but with highlighted edges.

- Shades of Gray visual style displays the color of all objects as shades of gray.

- Sketchy visual style displays edge lines that extend beyond their limits with a jitter effect.

- X-Ray visual style displays objects as transparent, as if you can see through them.

The best way to learn the different effects of the visual styles is to experiment with them.

The View panel displays a visual style's effect on an object as soon as it's selected.

PRACTICE 1-2

Orbit, Steering Wheel, and Visual Styles

1. Start AutoCAD 2015.

2. Open Practice 1-2.dwg.

3. Use the Steering Wheel to navigate through the high-rise buildings.

4. Use different orbit commands to rotate the high-rise buildings to see them from different angles.

5. Use different visual styles to view the high-rise buildings in different views.

6. Save and close the file.

1.6 WHAT IS THE USER COORDINATE SYSTEM (UCS)?

To understand the User Coordinate System (UCS), it is essential to first have an understanding of AutoCAD's World Coordinate System (WCS). The WCS defines all points in XYZ space, with the horizontal X-axis and vertical Y-axis forming the XY plane, and the Z axis intersecting this plane perpendicularly. Working in 3D requires the manipulation of complex shapes and different construction planes. AutoCAD introduced the User Coordinate System (UCS) so that users could create the necessary planes for object creation.

The relationship between the X, Y, and Z axes remain 90°. See the UCS icon:

There are three ways to create a new UCS:

- Manipulating the UCS Icon

- UCS Command

- DUCS

1.7 CREATING A NEW UCS BY MANIPULATING THE UCS ICON

You can create a new UCS by manipulating the UCS icon. When you hover over the UCS icon it will turn gold. Clicking the UCS icon activates the process and changes it to the following:

As you can see, there is a grip at the 0,0,0 origin, and a sphere at the end of each axis. Hovering over the grip displays the following:

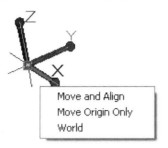

There are three options:

- Move and Align moves the UCS origin to a new location and aligns it with an existing face.

- Move Origin Only moves the UCS origin to a new location.

- World restores the WCS.

Hover over a sphere at the end of an axis, the following is displayed:

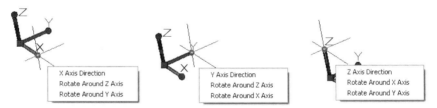

You can do two things from this menu:

- Redirect an axis (X, Y, or Z) by specifying a point

- Rotate around the other two axes

The UCS is mostly managed by using the grip at the 0,0,0 origin or the spheres at the end of the axes.

1.8.2 Origin Option

The Origin option creates a new UCS by moving the origin point from one point to another without modifying the orientation of the XYZ axes. To issue this command, select the **Origin** button:

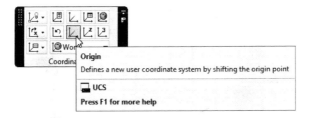

1.8.3 X/Y/Z Options

The X, Y, and Z options are similar and are therefore presented together. It is a simple concept: Fix one of the three axes and then rotate the other two to get a new UCS. How do you know whether the angle you have to input should be positive or negative? A simple rule provides guidance: Point your thumb of your right hand toward the positive portion of the fixed axis, and the curl of your four fingers will represent the positive direction. The Y option is used as an example below. To activate the Y command, select the **Y** button:

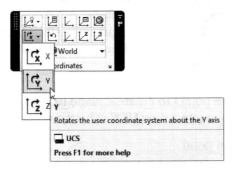

1.8.4 Named Option

You can name and save a UCS in AutoCAD so it can be used whenever you need it. After creating the UCS, go to the **View** tab, locate the **Coordinates** panel, and select the **Named** button:

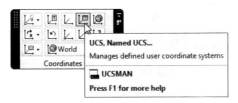

The following dialog box is displayed:

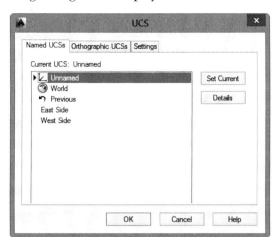

There is a list of saved UCSs under the **Named UCSs** tab, which also includes *Unnamed*, *World*, and *Previous*. Unnamed is the current UCS that was created before issuing this command. Click it to rename it, type the new name, and press [Enter]. Something similar to the following is displayed:

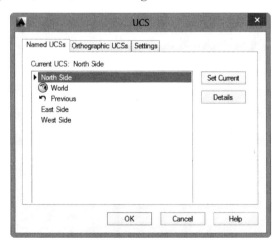

To end the command, click **OK**.

In the **Coordinates** panel there is a list of available preset UCSs that also displays any UCSs that you have saved, such as what is shown here:

1.8.5 Remaining Options

The previously mentioned options are most important. The remaining options are illustrated here:

- **World** option restores the WCS:

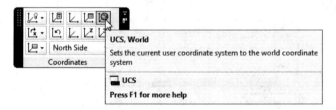

- **Z-Axis Vector** option creates a new UCS by specifying a new origin and a new Z-axis:

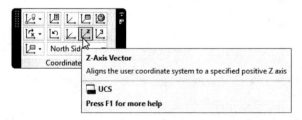

- **View** option creates a new UCS parallel to the screen (helpful if you want to input text for a 3D model):

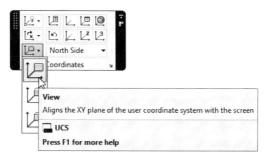

- **Object** option aligns the new UCS based on a selected object:

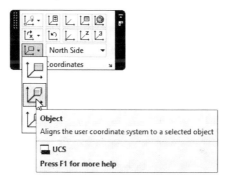

- **Face** option creates a new UCS by selecting the face of a solid:

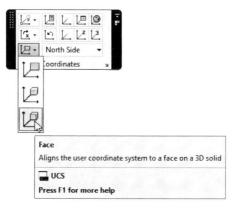

- **Previous** option restores the previous UCS:

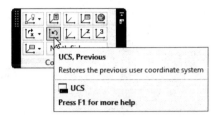

Hover over the UCS icon, right-click, and the following menu is displayed:

An alternative method to access all UCS commands is presented.

1.8.6 UCS Icon

The UCS icon is the XYZ symbol in the lower left corner of the screen. In the **Coordinates** panel, you can select if and where the UCS icon is shown. See the following:

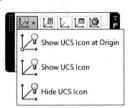

This list provides the following options:

- **Show UCS Icon at Origin** moves the UCS Icon to the current origin. This option is very handy if you are creating multiple UCSs.

- **Show UCS Icon** locates the UCS icon at the lower left corner of the screen.

- **Hide UCS Icon** removes the UCS icon from the screen. This is not recommended unless you have a specific purpose for doing so.

The **Properties** button found in the **Coordinates** panel provides control of the shape, color, and size of the UCS icon:

The following dialog box is displayed:

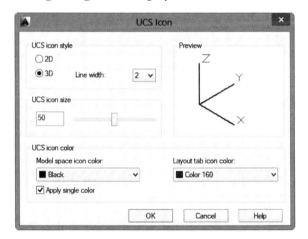

In the dialog box you can change the following:

- UCS Icon Style: 2D or 3D UCS icon, and line width

- UCS Icon Size

- UCS Icon Color

1.9 DUCS COMMAND

The Dynamic UCS (DUCS) works with other drafting and modifying commands in a simple and straightforward manner:

- Activate DUCS from the status bar:

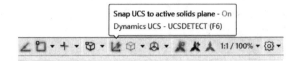

- Start a drafting command (line, circle, box, cylinder, etc.).

- Specify the first point by hovering over the face on which you want to draw. It is highlighted to indicate that AutoCAD has selected this face as the plane on which to draw. Specify the point and complete the command.

- When you are done, the operative UCS prior to the command is restored.

Using DUCS has advantages and disadvantages. The main advantages are that it is fast and simple. The disadvantages are as follows:

- It requires an object to be previously created.

- It is lost after completing the drafting command.

- It is limited to the faces of the selected object.

It is recommended to master all three UCS methods, and to utilize DUCS when appropriate.

1.10 TWO FACTS ABOUT UCS AND DUCS

Two important facts to consider when using user coordinate systems:

- The ViewCube shows views in relation to WCS or the current UCS. We recommend that you check the ViewCube to understand which UCS AutoCAD is currently using, and verify that the view is correct. See the following illustration:

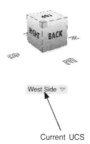

West Side

Current UCS

- The second fact is that OSNAP has priority over UCS, so it is possible to use OSNAP to pick points that are not in the current UCS. This leads to OSNAP and UCS contradiction. Users should be mindful of this, and be extra cautious when defining the UCS and then picking points not in the current UCS.

PRACTICE 1-3

User Coordinate System

1. Start AutoCAD 2015.

2. Open Practice 1-3.dwg.

3. Use any of the UCS creation methods to draw the following shapes on the existing model:

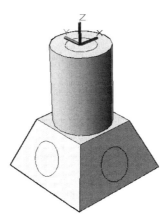

4. Save and close the file.

CHAPTER REVIEW

1. There are three Orbit commands.

    a. True

    b. False

2. _____ is a template to help you utilize the 3D environment.

3. One of the following is not an option in the UCS command:

    a. Face

    b. Rotate around X

    c. Realistic

    d. 3-Point

4. Using the ViewCube, you can change the visual style.

    a. True

    b. False

5. Which of the following is not an option of the Steering Wheel?

    a. Walk

    b. Center

    c. Conceptual

    d. Up/Down

6. _____ is a workspace that displays all ribbons related to 3D.

CHAPTER REVIEW ANSWERS

1. a

3. c

5. c

CHAPTER 2

CREATING SOLIDS

In This Chapter

- Basic Solid Shapes
- Manipulating Solids
- Presspull Command
- 3D Object Snaps
- Subobjects and Gizmo

2.1 INTRODUCTION TO SOLIDS

Solids are 3D objects that contain faces, edges, vertices, and contain substances. All of this information makes this object a very accurate 3D representation.

AutoCAD can be used to create many different types of solid shapes, including:

- Seven basic solid shapes
- Complex solid shapes using Boolean operations (union, subtract, and intersect)
- Complex solid shapes by converting 2D objects using the Presspull command
- Complex solid shapes by using Subobjects and Gizmo

2.2 CREATING SOLIDS USING BASIC SHAPES

AutoCAD provides seven basic solid shapes. Basic shapes are the first step in creating more complex shapes. To find these commands, go to the **Home** tab, locate the **Modeling** panel, and select one of the following:

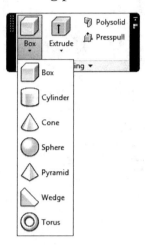

The commands are explained in the following sections.

2.2.1 Box Command

This command enables the creation of a box or cube. Select the **Box** command and the following prompts are displayed:

```
Specify first corner or [Center]:
Specify other corner or [Cube/Length]:
Specify height or [2Point]:
```

To draw a box in AutoCAD, you have to first define the base, then the height. The base is drawn on the current XY plane. Here are some variations:

- Specify the base by typing the coordinates of the first corner and the coordinates of the opposite corner.

- Specify the base by typing the coordinates of the first corner. To specify the opposite corner, specify the length and height by using the [Tab] key and typing.

- Specify the base by specifying the first corner, then specify Length (in the X axis direction) and Width (in the Y axis direction).

- Specify the base by specifying the first corner, then select the Cube option. AutoCAD asks you to input one dimension for all sides (length, width, and height).

- Specify the base by specifying the Center point and one of the corners of the base.

- After specifying the base, specify the height by typing it, using the mouse, or by specifying two points by clicking (this is a good method if you have an existing object).

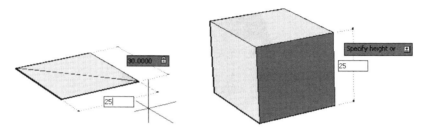

NOTE   *You can use dynamic input to input any of the distances needed to complete a command.*

2.2.2 Cylinder Command

This command enables the creation of a cylinder. Select the **Cylinder** command and the following prompts are displayed:

```
Specify center point of base or
[3P/2P/Ttr/Elliptical]:
Specify base radius or [Diameter]:
Specify height or [2Point/Axis endpoint]:
```

To draw a cylinder in AutoCAD, define the base, then the height. Here are some variations:

- The base could be a circle or an ellipse. The options to draw a circle or an ellipse are the same as they are with 2D prompts. The base is drawn on the current XY plane.

- After defining the base, specify the height by typing, by using the mouse, or by using the 2 Points method (using the height of an existing object).

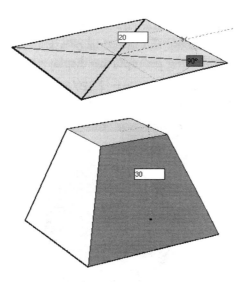

2.2.6 Wedge Command

This command enables the creation of a wedge. Select the **Wedge** command and the following prompts are displayed:

```
Specify first corner or [Center]:
Specify other corner or [Cube/Length]:
Specify height or [2Point]:
```

Wedge command prompts are the same as the Box command prompts. In fact, a wedge is a section of a box, as you first define the base and then define the height—the wedge is formed by the YZ plane toward the X axis:

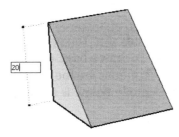

2.2.7 Torus Command

This command enables the creation of a torus. Select the **Torus** command and the following prompts are displayed:

```
Specify center point or [3P/2P/Ttr]:
Specify radius or [Diameter]:
Specify tube radius or [2Point/Diameter]:
```

Specify the center and radius (or diameter) of the torus, and then specify the radius (or diameter) of the tube. See the following illustration:

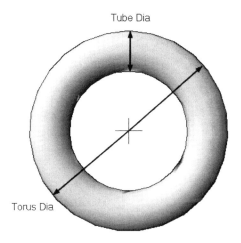

2.3 EDITING BASIC SOLID SHAPES USING GRIPS

Click one of the seven basic shapes and grips appear at certain locations, depending on the shape. These grips allow the modification of the dimensions of the basic shape in all directions. The grips for a box are shown here:

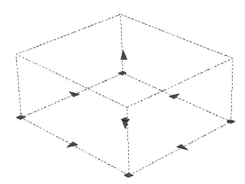

Cylinder grips:

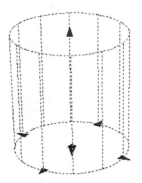

Cone grips:

Sphere grips:

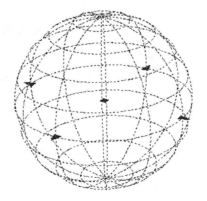

Pyramid grips:

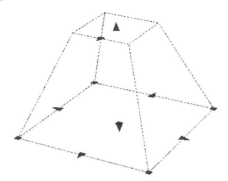

Wedge grips:

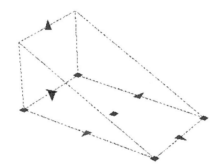

Torus grips:

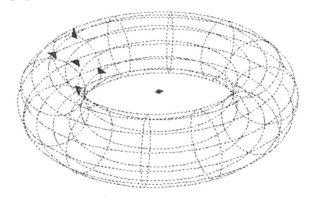

Grips are rectangular and triangular (arrows). The rectangular grips adjust two dimensions at the same time, as shown in the following illustration:

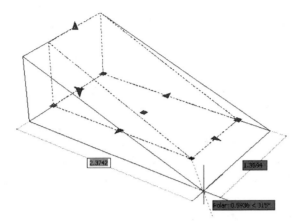

The arrows change only one dimension, in one direction, as illustrated in the following:

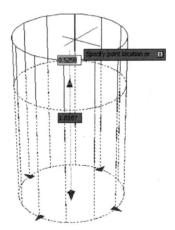

Dimensions can be entered by typing them in. To input more than one dimension, use the [Tab] key to jump from one number to another.

Grips can reveal the dimensions of edges of the shapes. See the following illustrations:

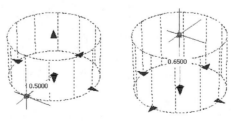

8. Using any of the editing methods learned, change the radius of the top cylinder to 5.

9. Change the width of the box to 5.

10. Save and close the file.

2.6 BOOLEAN FUNCTIONS

This is the first method to create complex solid shapes. The three Boolean functions are: Union, Subtract, and Intersect. To issue these commands, go to the **Home** tab, locate the **Solid Editing** panel, and find the three buttons on the left, as shown below:

2.6.1 Union Command

This command enables the union of all selected solids and surfaces in a single object. Select your objects in any order and press [Enter]. See the following example:

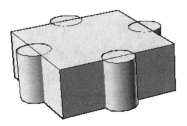

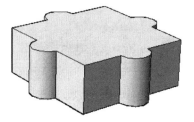

Before Union   After Union

2.6.2 Subtract Command

This command enables volume to be deducted from the existing volume of a solid. Select the entire existing object volume, press [Enter], then pick the object volume to be subtracted, and press [Enter] to execute the command. See the following example:

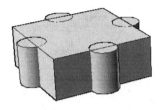

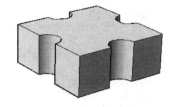

Before Subtract   After Subtract

2.6.3 Intersect Command

This command finds and creates the common volume of two or more solids. Solids can be picked in any order. See the following example:

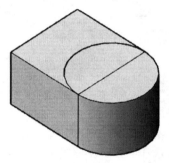

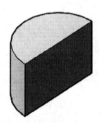

Before Intersect   After Intersect

2.6.4 Solid History

The big question is, "Does AutoCAD remember the original basic shapes that have contributed to this complex solid?" The answer to this question is, Yes!

To successfully record the history of a complex solid shape, follow these steps:

- Before creating the complex solid shape, go to the **Solid** tab, locate the **Primitive** panel, and select the **Solid History** button (if not already selected):

- Create your complex solid shape.

- Select the complex solid shape and activate the **Properties** palette. Under **Solid History**, make sure that **History = Record**. See the following illustration:

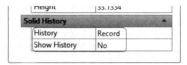

If you hold the [Ctrl] key and hover over any part of your complex solid, the original objects are highlighted. You can then click on them to select them and display their grips, enabling editing of their geometrical values. See the following illustration:

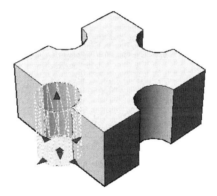

In **Properties**, if you set **Show History = Yes**, all solids that have contributed to the creation of the complex solid are displayed. Unfortunately, you cannot edit them in this state. To edit them, hold the [Ctrl] key and hover, as explained in the previous step. See the following illustration:

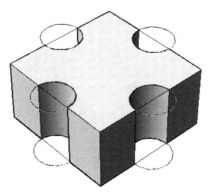

2.7 PRESSPULL COMMAND

The Presspull command is a very handy command that enables you to create a 3D solid on the fly. You can perform one of the following procedures:

- Presspull can offset an existing face of a 3D solid. Depending on the movement of the mouse, the 3D solid will increase or decrease.

- After drawing a 2D shape on a face of a 3D solid, the Presspull command enables the creation of a 3D solid, and then unions or subtracts it from the existing 3D solid.

- From a closed 2D shape, Presspull can create a 3D solid.

- From an open 2D shape, Presspull can create a 3D surface (discussed further in Chapter 4).

The Presspull command has a **Multiple** option (hold down the [Shift] key) that provides the ability to make several Presspull movements at the same time.

To issue this command, go to the **Home** tab, locate the **Modeling** panel, and select the **Presspull** button:

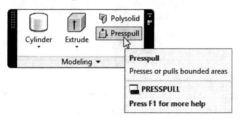

The following prompts are displayed:

```
Select object or bounded area:
Specify extrusion height or [Multiple]:
```

The command repeats these two prompts until you end it by pressing [Enter].

Here are examples of the four cases:

- Using the Presspull command to offset a 3D solid face:

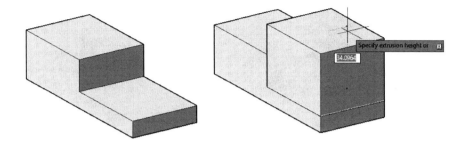

- After drawing a 2D object on a 3D solid face, press or pull the 2D shape to add or remove volume from the total 3D solid:

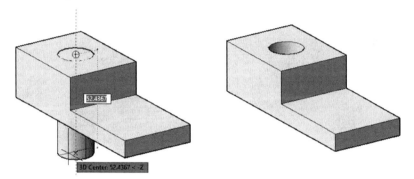

- Press or pull a closed 2D object to produce a 3D solid:

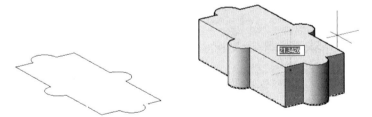

- Press or pull an open 2D object to produce a 3D surface:

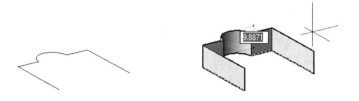

Using the Multiple option (hold the [Shift] key), you can select several objects (open 2D, closed 2D, face of 3D solid) and press or pull them at the same time.

You can also use the [Ctrl] key to control the output of the command. See the following example:

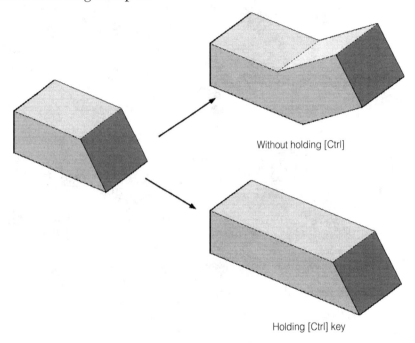

Without holding [Ctrl]

Holding [Ctrl] key

In the upper result, without holding the [Ctrl] key, the Presspull has offset the face perpendicular to the existing face. In the lower result, holding the [Ctrl] key, the Presspull has offset the face to follow the taper angle.

ON THE CD

PRACTICE 2-2

Boolean Functions and Presspull

1. Start AutoCAD 2015.

2. Open Practice 2-2.dwg.

3. Select the big box, right-click, and select Properties. Change the value of History to Record. Close the Properties palette.

4. Subtract the upper cylinder from the box.

5. Subtract the lower cylinder from the box (use two different commands).

6. Union the four boxes at the edges with the big box.

7. Hold the [Ctrl] key and click the upper cylinder. When the grips show up, change the radius to be 30 (add 10 units to the current radius).

8. Using the same technique, make the lower cylinder 10 units (remove 10 units from the current radius).

9. Using the Presspull command, pull the two upper sides up by 20 units.

10. Using Presspull command, press the front side to the inside by 15 units.

11. You will achieve something similar to the following:

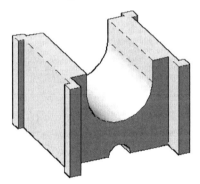

12. Thaw the 2D Objects layer.

13. Using the Presspull command, pull the closed shape at the XY plane up by 60 units.

14. Using Presspull, press the four circles at the edge to remove them from the volume of the shape (you may need to use Subtract).

15. Using the Presspull command, pull the two rectangles on the left side to the outside with distance = 15.

16. Using the Presspull command, pull the open shape up by 60 units.

17. Select the new shape. What type of shape is it? _____

18. This is the final shape:

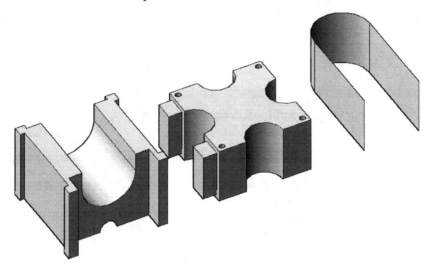

19. Save and close the file.

2.8 SUBOBJECTS AND GIZMO

A Subobject is any part of a solid; it can be a vertex, an edge, or a face. Gizmo is a gadget that will help you Move, Rotate, or Scale the selected vertices, edges, or faces. Both are available in the **Home** tab, on the **Selection** panel:

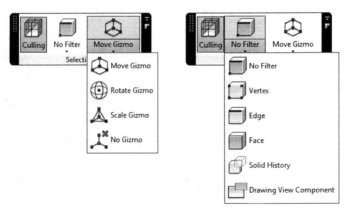

You can specify the filter and gizmo using the Status bar (excluding the No Filter and No Gizmo options), as shown here:

Select the desired type of filter and your cursor will change to the shape below (this cursor is for selecting vertices). Now when you select, you will only select the chosen subobject:

Below are examples of selecting subobjects:

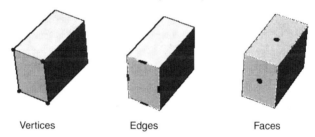

Vertices   Edges   Faces

There are three things to consider when selecting. These are:

2.8.1 Culling

The Culling feature is a very useful tool that allows you to select only the subobjects that are visible from the current viewpoint. To switch it on or off, go to the **Home** tab, locate the **Selection** panel, and click the **Culling** button:

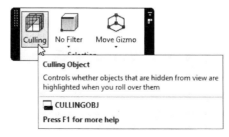

If the Culling button is off, all subobjects will be selected regardless of whether or not they are visible from the current viewpoint. See the following two examples:

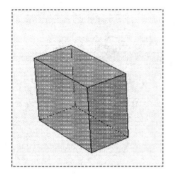

Culling is ON

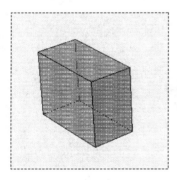
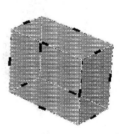

Culling is OFF

2.8.2 Selection Cycling

When you click on a face to select it, there may be other faces behind it. AutoCAD lets you select exactly what you want to select. It shows you a small window (the **Selection**) that includes all objects that exist in the same location of your click. This small window displays one object after another until you click the desired object. This feature is called **Selection Cycling**. To activate it, go to the **Drafting Settings** dialog box (used for Snap, Grid, Osnap, etc.), and select the **Selection Cycling** tab, as shown here:

For example, assume we have a pyramid, and we clicked the top face, as shown here:

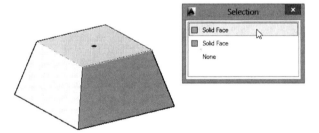

AutoCAD selects the face nearest to your click, but lets you to select the face behind it as well, and lets you decide. See the following illustration:

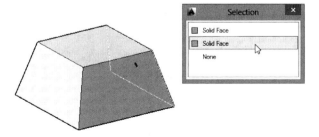

This applies to edges as well.

2.8.3 Selecting Preset Views

Using a Preset view lets you select multiple subobjects quickly and accurately. The following steps illustrate this concept:

- Click **Culling** off.

- Using the ViewCube, go to **Top** view.

- From the Subobjects panel, select **Edge**.

- Without issuing any command, use <u>Window</u> to contain all edges on the right and left sides of the box, as shown here:

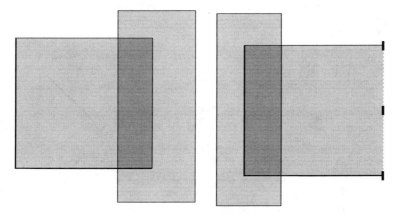

- Using the ViewCube, change to any 3D preset view and you will see something similar to the following:

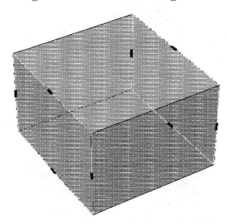

2.8.4 Gizmo

After selecting the desired subobjects, you can choose the gizmo to change the subobjects. There are three gizmos available:

- Move Gizmo:

- Rotate Gizmo:

- Scale Gizmo:

By default, the gizmo will go to the last selected subobject. To move the gizmo from the last selected subobject to another, move your cursor (without clicking) to the desired subobject, stay for couple of seconds, the gizmo will follow. Right-click displays the following menu:

This menu provides the following options:

- Change the Gizmo

- Set Constraints (this sets the axis or plane to be used)

- Relocate, Align, and Customize Gizmo

The Move gizmo consists of the three axes, enabling the movement of selected subobjects along one of them (namely X, Y, or Z), or along a plane of two of them (XY, YZ, XZ). To do this, hover over one of three axes until it is highlighted (it will become golden), then click and move. You can use the right-click menu to set the desired axis using the **Constraint** option. See the following illustration:

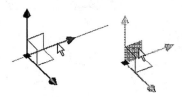

The Rotate gizmo consists of three 3D circles. To rotate around one of the three axes, move your cursor to the desired circle and pause for couple of seconds. When it is highlighted (it will become golden), click and rotate. You can use the right-click menu to set the desired axis using the Constraint option. See the following illustration:

The Scale gizmo consists of three axes. Select a plane (XY, YZ, or XZ), or space (XYZ), move your cursor to the desired plane and pause for couple of seconds until the plane is highlighted (it will become golden), click, then scale. You can use the right-click menu to set the desired axis using the Constraint option. See the following illustration:

The gizmo does not only act on subobjects, but also on solid shapes. Follow the following steps:

- Make sure the Subobject = No Filter.

- Make sure Culling is off (so you can see all grips).

- When you click a solid shape, gizmo appears at the center of the shape; this is your base point. If you want to use other base points, relocate the gizmo and select another grip as a new base point.

- Start and complete the desired command.

See the following illustrations:

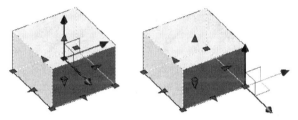

PRACTICE 2-3

Subobjects and Gizmo

1. Start AutoCAD 2015.

2. Open Practice 2-3.dwg.

3. Change Subobjects to Face, and move the upper face up by 5 units.

4. Using the same face, rotate it around X (the red circle) 15°.

5. Select the edges as shown, and move them outward 5 units:

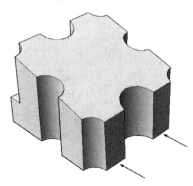

6. Rotate the shape to see the curved face at the back.

7. Select the curved face, and scale it by a factor of 1.5.

8. Select the two vertices as shown, and move them downward by 5 units:

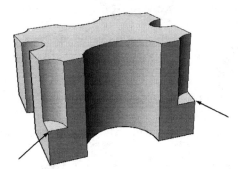

9. Save and close the file.

PROJECT 2-1-M

ON THE CD

Mechanical Project Using Metric Units

1. Start AutoCAD 2015.

2. Open Project 2-1-M.dwg.

3. Draw the 3D solid shape using the following information:

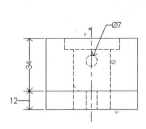

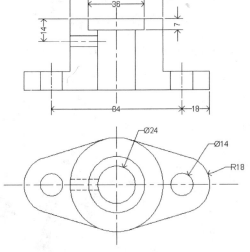

4. The final shape in 3D should look like the following:

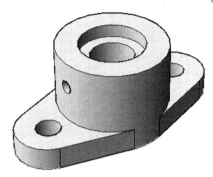

5. Save and close the file.

 PROJECT 2-1-I

ON THE CD

Mechanical Project Using Imperial Units

1. Start AutoCAD 2015.

2. Open Project 2-1-I.dwg.

3. Draw the 3D solid shape using the following information:

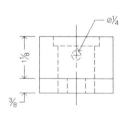

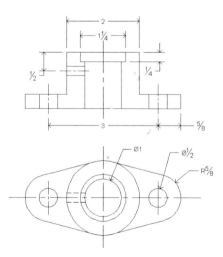

4. The final shape in 3D should look like the following:

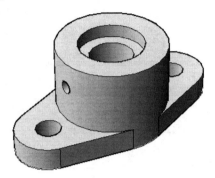

5. Save and close the file.

PROJECT 2-2-M

Architectural Project Using Metric Units

1. Start AutoCAD 2015.

2. Open Project 2-2-M.dwg.

3. Make sure the visual style is 2D Wireframe.

4. Create a new layer and call it 3D Wall, set the color to 9, and make it current.

5. Using the Presspull command, pull the outside walls up by 3,000 units, making sure that the door and window openings remain.

6. Create a new layer and call it 3D Door, set the color to 8, and make it current.

7. To draw the outer door jamb, explode the 2D Door block. Using Presspull, pull the two rectangles that represent the jamb up by 2,000 units.

8. To close the frame, draw a box with a base that is the two ends of the two boxes you just pulled, with height = 50.

9. Using the Union command, union the frame to be one piece.

10. To draw the door (it will be open) erase the arc and increase the length of the inside line by 1 unit (make the two line lengths = 800), then draw a line closing the shape. Using the Presspull command, pull the rectangle up by 2,000 units, creating the door.

11. Make the 3D Wall layer current.

12. To close the gap between the door and the wall, draw a box using the object snaps. Then, using the Union command, union the wall to be one piece.

13. Move the 2D block of the window on the right of the door.

14. Using Presspull, pull the rectangle at the location of the 2D window block up by 500 units.

15. Create a new layer and name it 3D Window, set its color to 8, and make it current.

16. Explode the 2D block of the window and erase the inner three lines. Presspull the two rectangles up by 1,900 units. Create two boxes at the two ends with a height of 50 to close the frame.

17. Union the four boxes to create one unit.

18. Move it to the opening.

19. Make layer 3D Wall current, and create a box to close the gap as you did in step 12.

20. Copy the three parts to all the windows with the same size.

21. Do the same thing with the big windows, but using a window height of 1,500 (1,400 + 50 + 50), and the distance from ground to the beginning of the window (Sill height) as 1,000.

22. Using the Union command, union all the walls.

23. Freeze layers A-Wall, A-Door, and A-Window.

24. Create a new layer and call it 3D Roof, set its color to 8, and make it current.

25. Create a wedge with a base of 2,350 x 11,400, and height = 2,000.

26. Mirror it around its longest dimension.

27. Union both parts.

28. Move the gable at the top of the wall of the entrance.

29. To create the second gable, create a wedge with a base 3,675 x 7,600, and height = 2,000.

30. Mirror it using the longest dimension, then union them.

31. Rotate them using the gizmo so they fit the other part of the roof.

32. Move the gable to the top of the walls.

33. To make the two gables close right. Change the subobject to be the vertex, then select the top vertex on the right, as in the following:

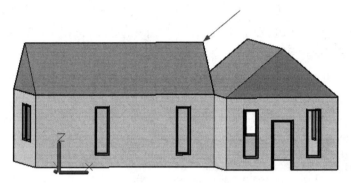

34. Using the Move gizmo, move it until it touches the other gable (you may need to use the Perpendicular object snap).

35. It should look like the following:

36. Create a new layer and call it 3D Base, set its color to 8, and make it current.

37. Look at the model from below.

38. Using object snaps of the model, create a box that covers the entire area of the house, reducing the height by 50.

39. Select the base to show the grips. Using the grips from the side of the entrance, set the dimension ***to be*** 20,000 units.

40. From the other three sides, increase the dimension by 3,000 units.

41. Create a new layer and call it 3D Stairs, set the color to 8, and make it current.

42. In an empty place, create a box 10,000 x 2,750, with height = 200 in the negative.

43. Move it from the midpoint of the top edge to the midpoint of the base on the entrance side.

44. Copy it four times to create a staircase that goes down.

45. Create a box 18,300 x 15,000 height = 50 in the negative.

46. Move it from its midpoint to the midpoint of the last step.

47. Create a new layer and call it 3D Entrance, set the color to 8, and make it current.

48. Start the Cylinder command, set the center point to be the midpoint of the right edge of the last box drawn, and set the radius to 500 with height = 6,000.

49. Move it to the inside by 500 using gizmo.

50. Copy it to the other side.

51. Using 3D Osnap, copy it to the center of the face to make a total of three cylinders.

52. In an empty space, draw a box 8,650 x 5,250, with height = 900.

53. Using DUCS, draw an arc on the upper face using Start-Center-End where the start is the bottom right endpoint, and the center is the midpoint of lower edge, and the endpoint of the arc is the other endpoint of the edge.

54. Using Presspull, press the arc downward to subtract it from the shape.

55. Using the gizmo, rotate the shape by 90°.

56. Move the shape from the midpoint to the center of the cylinder.

57. Copy it to the other side.

58. Union both shapes.

59. This is the final shape.

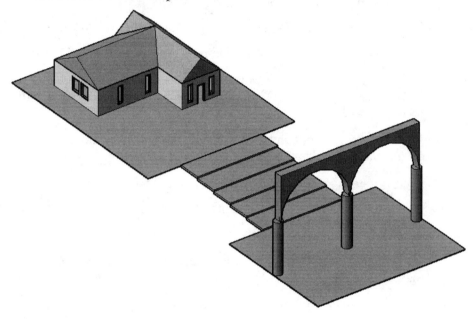

60. Save and close the file.

PROJECT 2-2-I

Architectural Project Using Imperial Units

1. Start AutoCAD 2015.

2. Open Project 2-2-I.dwg.

3. Make sure the visual style is 2D Wireframe.

4. Create a new layer and call it 3D Wall, set its color to 9, and make it current.

5. Using the Presspull command, pull the outside walls up by 120 inches (or 10 ft), making sure that the door and window openings remian.

6. Create a new layer and call it 3D Door, set the color to 8, and make it current.

7. To draw the outer door jamb, explode the 2D Door block. Using Press-pull, pull the two rectangles that represent the jamb up by 80 inches.

8. To close the frame, draw a box with a base that is the two ends of the two boxes you just pulled, with height = 2 inch.

9. Using the Union command, union the frame to be one piece.

10. To draw the door (it will be open) erase the arc, and then draw a line to close the shape. Using the Presspull command, pull the rectangle by 80 inches up, creating the door.

11. Make the 3D Wall layer current.

12. To close the gap between the door and the wall, draw a box using the object snaps. Then, using the Union command, union the wall to be one piece.

13. Move the 2D block of the window on the right of the door.

14. Using Presspull, pull the rectangle at the location of the 2D window block up by 20 inches.

15. Create a new layer and name it 3D Window, set its color to 8, and make it current.

16. Explode the 2D block of the window and erase the inner three lines. Presspull the two rectangles up by 76 inches. Create two boxes at the two ends with a height of 2 inches to close the frame.

17. Union the four boxes to create one unit.

18. Move it to the opening.

19. Make the 3D Wall layer current, and create a box to close the gap as you did in step 12.

20. Copy the three parts to all windows with the same size.

21. Do the same thing with the big window, but using a window height of 60 inches (56 + 2 + 2), and a distance from the ground to the beginning of the window (Sill height) of 40 inches.

22. Using the Union command, union all walls.

23. Freeze layers A-Wall, A-Door, and A-Window.

24. Create a new layer and call it 3D Roof, set its color to 8, and make it current.

25. Create a wedge with a base of 7'10" x 38', and height = 80 inches.

26. Mirror it around its longest dimension.

27. Union both parts.

28. Move the gable at the top of the wall of the entrance.

29. To create the second gable, create a wedge with a base of 12'3" x 25'4" with height = 80.

30. Mirror it using the longest dimension, then union them.

31. Rotate them using the gizmo to make them fit the other part of the roof.

32. Move the gable to the top of the walls.

33. To make the two gables close right, change the subobject to be the vertex, then select the top vertex on the right, as in the following:

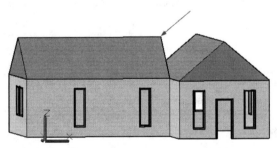

34. Using the Move gizmo, move it until it touches the other gable (you may need to use the Perpendicular object snap).

35. It should look like the following:

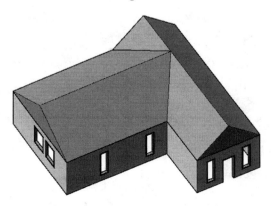

36. Create a new layer called 3D Base, set its color to 8, and make it current.

37. Look at the model from below.

38. Using object snaps of the model, create a box that covers all the area of the house with height reduced by 2 inches.

39. Select the base to show the grips. Using the grips from the side of the entrance, set the dimension ___to be___ 67'.

40. From the other three sides, increase the dimension by 120 inches.

41. Create a new layer called 3D Stairs, set the color to 8, and make it current.

42. In an empty place, create a box 33' x 9' with height = 8 inches in the negative.

43. Move it from the midpoint of the top edge to the midpoint of the base on the entrance side.

44. Copy it four times to create a staircase that goes down.

45. Create a box that is 61' x 50' and height = 2 inches in the negative.

46. Move it from its midpoint to the midpoint of the last step.

47. Create a new layer called 3D Entrance, set the color to 8, and make it current.

48. Start the Cylinder command, set the center point to be the midpoint of the right edge of the last box drawn, and set the radius to be 20 inches with height = 20'.

49. Move it to the inside by 20 inches using the gizmo.

50. Copy it to the other side.

51. Using 3D Osnap, copy it to the center of the face to make a total of three cylinders.

52. In an empty space, draw a box 28'10" x 17'6", with height = 36 inches.

53. Using DUCS, draw an arc on the upper face using Start-Center-End, where the start is the bottom right endpoint, the center is the midpoint of lower edge, and the endpoint of the arc is the other endpoint of the edge.

54. Using Presspull, press the arc downward to subtract it from the shape.

55. Using the gizmo, rotate the shape by 90°.

56. Move the shape from the midpoint to the center of the cylinder.

57. Copy it to the other side.

58. Union both shapes.

59. This is the final shape.

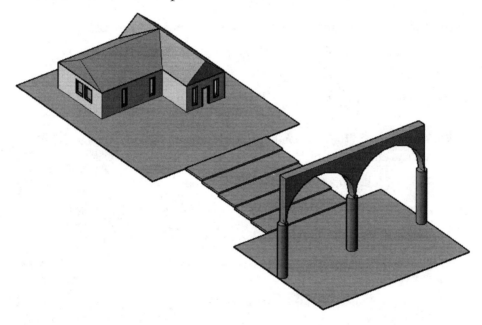

60. Save and close the file.

PROJECT 2-3-M

Mechanical Project Using Metric Units

1. Start AutoCAD 2015.

2. Open Project 2-3-M.dwg.

3. Draw the 3D solid shape using the following information:

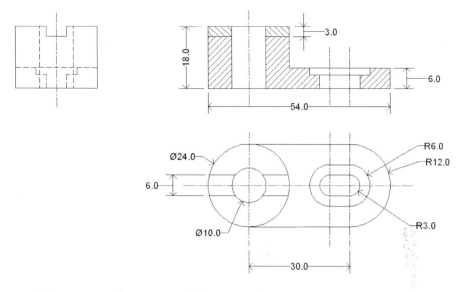

4. The missing dimension of the top right image is 2.

5. The final shape in 3D should look like the following:

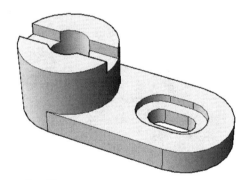

6. Save and close the file.

PROJECT 2-3-I

Mechanical Project Using Imperial Units

1. Start AutoCAD 2015.

2. Open Project 2-3-I.dwg.

3. Draw the 3D solid shape using the following information:

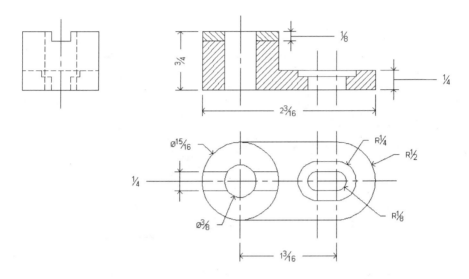

4. The missing dimension of the top right image is 1/12.

5. The final shape in 3D should look like the following:

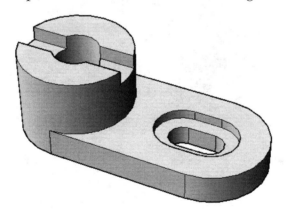

6. Save and close the file.

NOTES

CHAPTER REVIEW

1. Boolean operations are three commands.

   a. True

   b. False

2. When you issue the Presspull command, you can press or pull any number of 2D objects in the same command.

   a. True

   b. False

3. Vertex, Edge, and Face, are _____ of a solid.

4. One of the following statements is not true:

   a. Solid History should be ON before the creation of the complex solid shape if you want to edit individual solid objects.

   b. If Culling = on, it helps you to select all edges that are seen or not seen from the current viewpoint.

   c. Using preset views helps you select subobjects faster.

   d. There are three gizmo commands.

5. You can move vertex, edges, and faces parallel to an axis or in a plane.

   a. True

   b. False

6. 3D Center is _____.

7. Which feature in AutoCAD helps you select a face behind the face you clicked?

   a. Culling

   b. Selecting using Window

   c. Selection cycling

   d. 3D Osnap

8. Selection cycling and 3D Osnap can be controlled from:

    a. Solid tab, Selection panel

    b. Home tab, Selection panel

    c. Status bar

    d. ViewCube

CHAPTER REVIEW ANSWERS

    1. a

    3. Subobjects

    5. a

    7. c

CHAPTER 3

CREATING MESHES

In This Chapter

- Meshes: Basic Shapes
- Meshes: Subobjects and Gizmo
- Creating Meshes from 2D objects
- Converting, Smoothing, Refining, and Creasing
- Face-Editing Commands

3.1 MESHES

Meshes were first introduced as 3D objects in AutoCAD 2010. They look like solids, but don't have mass or volume. Unlike solids, Meshes can form irregular shapes. Autodesk therefore calls this *Free Form* design.

Meshes consist of faces specified by the user. These faces consist of facets, which AutoCAD manipulates to create a smoother mesh.

Facets are controlled by AutoCAD; users cannot edit them.

3.2 BASIC MESH SHAPES

There are seven basic mesh shapes, just like solids. All commands can be found under the **Mesh** tab, on the **Primitives** panel:

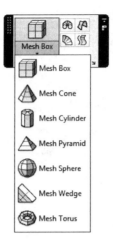

Drawing a basic mesh shape is the same as it is for a basic solid shape, so your experience can be transferred to meshes without hassle. See the following illustration to differentiate between a solid box and a mesh box:

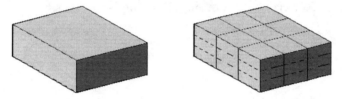

To show the internal edges, set the system variable **vsedges** to 1.

How do you control the number of faces in each direction? Meshes have default tessellation divisions that can be changed by going to the **Mesh** tab, locating the **Primitives** panel, and clicking the **Mesh Primitive Options** dialog box:

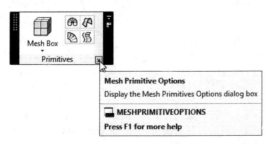

The following dialog box is displayed:

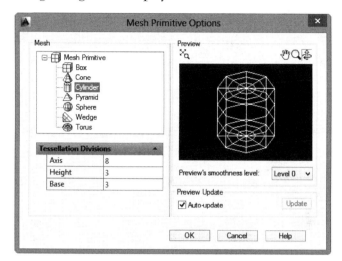

Users should pick the shape they want to alter, and then set the tessellation (number of faces) in all directions, depending on the shape.

3.3 MANIPULATING MESHES USING SUBOBJECTS AND GIZMO

Subobjects and gizmos provide the ability to manipulate any basic mesh shape to create an irregular shape by using the three gizmo commands.

See the following example:

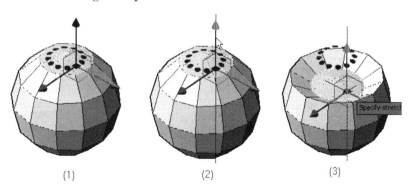

3.4 INCREASING AND DECREASING MESH SMOOTHNESS

A mesh can display four levels of smoothness, from None (no smoothness) to level 4 (smoothest). Use Quick Properties to increase and decrease the smoothness of a mesh. See the following illustration:

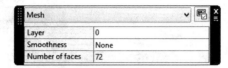

Alternatively, go to the **Mesh** tab, locate the **Mesh** panel, and select the **Smooth More** or **Smooth Less** button:

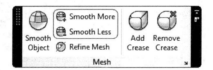

AutoCAD displays the following prompts (for Smooth More):

```
Select mesh objects to increase the smoothness
level:
Select mesh objects to increase the smoothness
level:
```

See the following illustration:

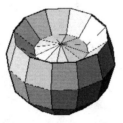

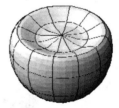

Smoothness = None   Smoothness = Level 2

PRACTICE 3-1

Basic Mesh Objects, Subobjects, and Gizmo

1. Start AutoCAD 2015.

2. Open Practice 3-1.dwg.

3. Go to Mesh Primitive Options, select Sphere, and make sure that Axis = 12 and Height = 6.

4. Change the visual style to Conceptual and set VSEDGES = 1.

5. Draw a mesh sphere with Radius = 20.

6. Change subobject = Face.

7. Select all top faces and move them down 15 units.

8. Select all bottom faces and move them up 15 units.

9. Change subobject = No Filter.

10. Select the mesh and change Smoothness = Level 2.

11. Change subobject = Edge.

12. Select the following edges:

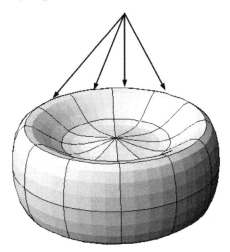

13. Move them upward 15 units.

14. You should have the following shape:

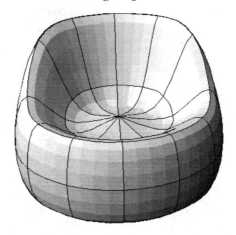

15. Save and close the file.

3.5 CONVERTING 2D OBJECTS TO MESHES

There are four commands that convert closed or open 2D shapes to meshes. These commands are:

- Revolved Surface
- Edge Surface
- Ruled Surface
- Tabulated Surface

To set the density of meshes generated by these four commands, use the system variables SURFTAB1 and SURFTAB2. Ruled surface and Tabulated surface commands are only affected by SURFTAB1, while the Revolved surface and Edge surface commands are affected by both SURFTAB1 and SURFTAB2.

3.5.1 Revolved Surface Command

This command creates a mesh from an open or closed 2D object by revolving it around an axis using an angle. To issue this command, go to the **Mesh** tab, locate the **Primitives** panel, and select the **Revolved Surface** button:

The following prompts are displayed:

```
Current wire frame density:  SURFTAB1=16
SURFTAB2=16
Select object to revolve:
Select object that defines the axis of revolution:
Specify start angle <0>:
Specify included angle (+=ccw, -=cw) <360>:
```

The first line presents the current values of SURFTAB1 and SURFTAB2, and is repeated with the three other commands.

The second prompt asks you to select one open or closed 2D object. The third prompt asks you to select an object that represents the axis of revolution. The fourth prompt asks you to specify the start angle, if other than 0 (zero). Finally, you are asked to specify the included angle, bearing in mind that CCW is positive.

The result is something similar to the following:

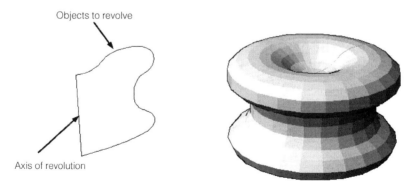

Objects to revolve

Axis of revolution

3.5.2 Edge Surface Command

This command creates a mesh from four open 2D shapes. For this command to be successful, the four objects should touch each other to form a closed four-edged shape. To issue this command, go to the **Mesh** tab, locate the **Primitives** panel, and select the **Edge Surface** button:

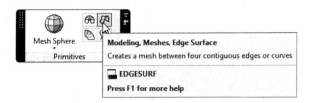

The following prompts are displayed:

```
Current wire frame density:  SURFTAB1=16
SURFTAB2=16
Select object 1 for surface edge:
Select object 2 for surface edge:
Select object 3 for surface edge:
Select object 4 for surface edge:
```

The first line presents the current values of SURFTAB1 and SURFTAB2. Select the four edges one by one. The result is something similar to the following:

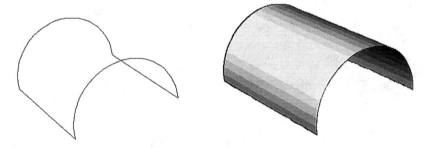

3.5.3 Ruled Surface Command

This command creates a mesh between two objects in one of the following variations:

- Two open 2D shapes
- Two closed 2D shapes
- Open shape with point
- Closed shape with point

To issue this command, go to the **Mesh** tab, locate the **Primitives** panel, and select the **Ruled Surface** button:

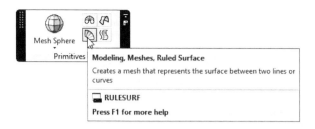

AutoCAD displays the following prompts:

```
Current wire frame density:  SURFTAB1=24
Select first defining curve:
Select second defining curve:
```

The first line presents the current value of SURFTAB1. The second and third prompts ask you to select whether the two defining curves are open, closed, or a point. See the following illustration:

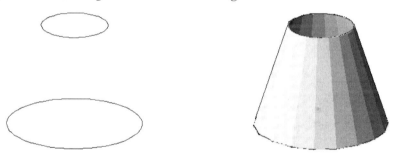

3.5.4 Tabulated Surface Command

This command creates a mesh using an open or closed 2D shape (path curve) and an object (direction vector). To issue this command, go to the **Mesh** tab, locate the **Primitives** panel, and select the **Tabulated Surface** button:

AutoCAD displays the following prompts:

```
Current wire frame density:  SURFTAB1=24
Select object for path curve:
Select object for direction vector:
```

The first line presents the current value of SURFTAB1. The second prompt asks you to select the path curve, and the third prompt asks you to select the direction vector. See the following illustration:

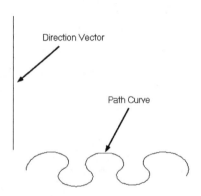

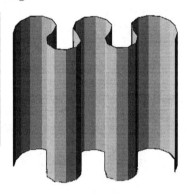

ON THE CD

PRACTICE 3-2

Converting 2D Objects to Meshes

1. Start AutoCAD 2015.

2. Open Practice 3-2.dwg.

3. Set both SURFTAB1 and SURFTAB2 to 24.

4. Change the visual style to Conceptual.

5. Using two arcs and the lines in the X direction, create an Edge surface.

6. Using the arcs and the lines in the Y direction, create a Rule surface.

7. Using the four circles and the vertical line, create a Tabulated surface.

8. Using the line and the polyline, create a Revolved surface.

9. You should have something similar to the following:

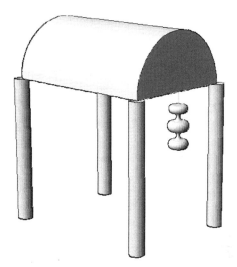

10. Save and close the file.

3.6 CONVERTING, REFINING, AND CREASING

This section discusses the three commands that allow you to:

- Convert legacy surfaces or solids to meshes
- Increase the number of faces in a mesh
- Add or remove creases to or from meshes

3.6.1 Converting Legacy Surfaces or Solids to Meshes

Prior to AutoCAD 2007, a type of 3D object known as surfaces existed. This type of object has no relation to the surfaces discussed in Chapter 4. If you open a file that contains objects of this type, and you want to convert them to meshes—or convert any solid to a mesh—this command is for you. To issue this command, go to the **Mesh** tab, locate the **Mesh** panel, and select the **Smooth Object** button:

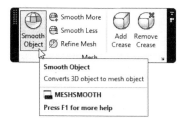

The following prompts are displayed:

```
Select objects to convert:
Select objects to convert:
```

Select the 3D objects you want to convert. When done, press [Enter] to execute the command. See the following illustration:

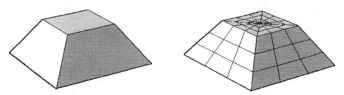

How do you control the result of the mesh? How many faces will be in different directions? Will it be smoothed or not? To answer these questions, go to the **Mesh** tab, locate the **Mesh** panel, and select the **Mesh Tessellation Options** button:

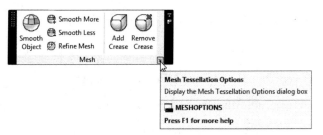

The following dialog box is displayed:

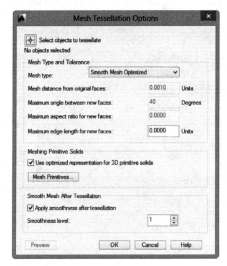

If you select **Smooth Mesh Optimized** for the **Mesh type**, many of the options are disabled. The remaining options are self-explanatory.

3.6.2 Refining Meshes

You can increase the number of faces on a mesh face, or in the whole mesh. The smoothness level should be 1 or more. To issue this command, go to the **Mesh** tab, locate the **Mesh** panel, and select the **Refine Mesh** button:

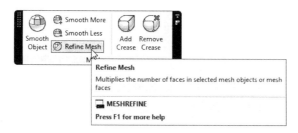

The following prompts are displayed:

```
Select mesh object or face subobjects to refine:
Select mesh object or face subobjects to refine:
```

Select a mesh or face subobject. If you try to refine a mesh with smoothness level = None, the following message is displayed:

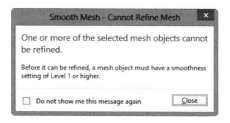

When done, press [Enter] to execute the command. See the following illustration:

 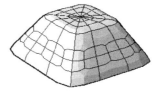

3.6.3 Add or Remove Crease

This command decreases smoothness; it adds texture to the faces of your mesh. To issue this command, go to the **Mesh** tab, locate the **Mesh** panel, and select the **Add Crease** or **Remove Crease** button:

The following prompts are displayed:

```
Select mesh subobjects to crease:
Select mesh subobjects to crease:
Specify crease value [Always] <Always>:
```

You can crease faces, edges, or vertices. The last prompt specifies the **crease value**, which is the highest smoothness level value that preserves the crease; a value greater than this value completely removes the crease. **Always** means the crease is always preserved. See the following illustration:

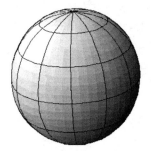 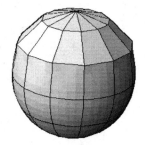

If you click a mesh, two new context panels are added to the *current* tab: **Smooth**, which includes the commands discussed above, and **Convert Mesh**, which is discussed later in this book.

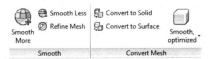

PRACTICE 3-3

Converting, Refining, and Creasing

1. Start AutoCAD 2015.

2. Open Practice 3-3.dwg.

3. Change Smoothness = Level 1.

4. Refine the two faces as shown below:

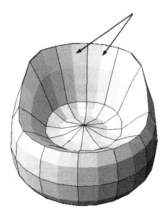

5. Add crease to the two faces behind the faces selected in the previous step.

6. Change Smoothness = Level 4.

7. Thaw the Legs layer.

8. Convert the four spheres to meshes.

9. This is the final result:

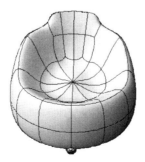

10. Save and close the file.

3.7.3 Merge Face

This command performs the opposite of Split face. It merges two or more adjacent mesh faces. To issue this command, go to the **Mesh** tab, locate the **Mesh Edit** panel, and select the **Merge Face** button:

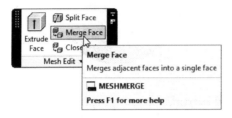

The following prompts are shown:

```
Select adjacent mesh faces to merge:
Select adjacent mesh faces to merge:
Select adjacent mesh faces to merge:
```

Select the faces to be merged, and press [Enter]. See the following illustration:

 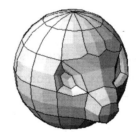

3.7.4 Closing a Hole in a Mesh

No command makes a hole in a mesh, as you simply select a face and press [Del]. To close a hole in a mesh, go to the **Mesh** tab, locate the **Mesh Edit** panel, and select the **Close Hole** button:

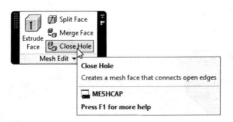

The following prompts are shown:

```
Select connecting mesh edges to create a new mesh
face:
Select connecting mesh edges to create a new mesh
face:
```

Select all edges of the hole (they should be on the same plane), and, when done, press [Enter] to execute the command. See the following illustration:

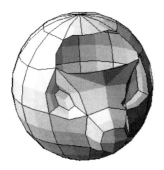 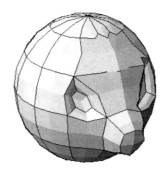

3.7.5 Collapse a Face or Edge

This command deletes a face or edge. It doesn't make a hole; it merges the vertices of the selected faces or edges, as if you are reducing the number of faces in the mesh. To issue this command, go to the **Mesh** tab, locate the **Mesh Edit** panel, and select the **Collapse Face or Edges** button:

The following prompts are shown:

```
Select mesh face or edge to collapse:
```

Select the desired edges or faces. See the following illustration:

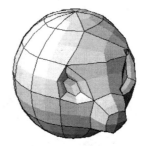

3.7.6 Spin Triangular Face

This command spins the shared edge of two triangular mesh faces. To issue this command, go to the **Mesh** tab, locate the **Mesh Edit** panel, and select the **Spin Triangular Face** button:

The following prompts are shown:

```
Select first triangular mesh face to spin:
Select second adjacent triangular mesh face to
spin:
```

Select the two triangular faces that share the same edge, and you will see something similar to the following:

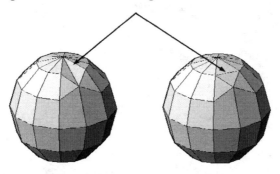

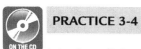

PRACTICE 3-4

Mesh Editing Commands

1. Start AutoCAD 2015.

2. Open Practice 3-4.dwg.

3. Split the two faces, as shown below, and extrude the face by 7 units:

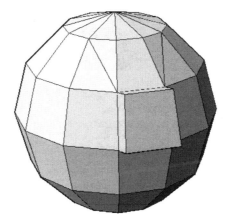

4. Split the face, as shown below, and extrude the faces by 2.5 units to the inside to create the following shape:

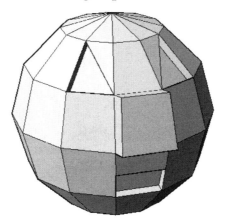

5. Collapse the edges shown below:

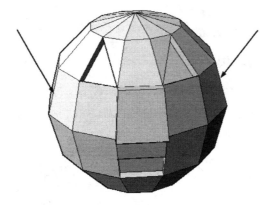

6. Change Smootheness = Level 4.

7. You will get the following shape:

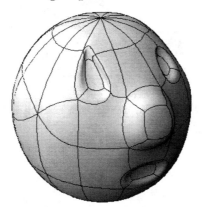

8. Save and close the file.

PROJECT 3-1-M

Project Using Metric Units

1. Start AutoCAD 2015.

2. Open Project 3-1-M.dwg.

3. Change the visual style to Conceptual (change VSEDGES = 1).

4. Start Mesh Primitives Options and set the following values for the Box: Length = 8, Width = 8, Height = 1.

5. Draw a mesh box 120 x 120 x 60.

6. Select all the faces as shown below:

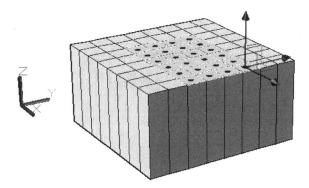

7. Using gizmo, move the faces downward by 32 units.

8. Select the edges as shown below:

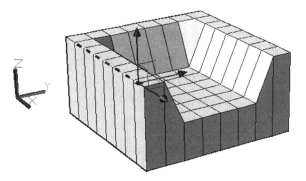

9. Move the edges toward the negative of Y-axis by 20 units.

10. Move the same edges downward 10 units.

11. Do the same for the other side.

12. Extrude faces to the outside by 12 units.

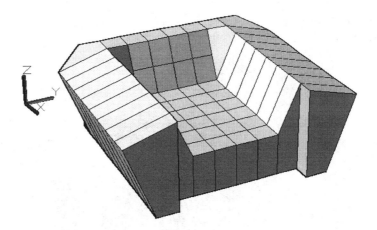

13. Merge the faces, as shown below, and move it downward 6 units.

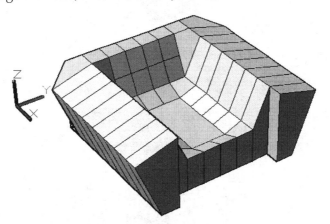

14. Move the faces upward 16 units, as shown below:

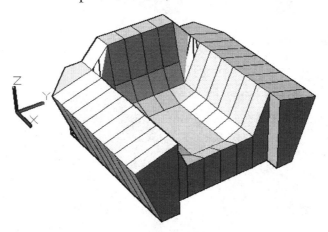

15. Change the smoothness of the shape to Level 4.

16. Select the faces, as shown below, and set the crease value to 1.

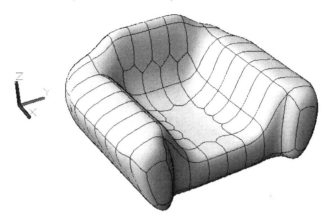

17. Save and close the file.

PROJECT 3-1-I

Project Using Imperial Units

1. Start AutoCAD 2015.

2. Open Project 3-1-I.dwg.

3. Change the visual style to Conceptual (change VSEDGES = 1).

4. Start Mesh Primitives Options and set the following values for the Box: Length = 8, Width = 8, Height = 1.

5. Draw a mesh box 48" x 48" x 24".

6. Select all the faces as shown below:

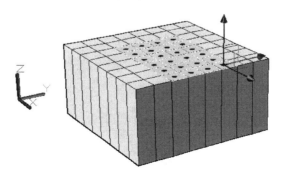

7. Using gizmo, move the faces downward 13".

8. Select the edges as shown below:

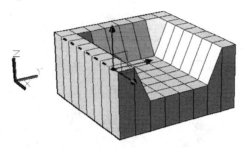

9. Move the edges toward the negative of Y-axis by 8".

10. Move the same edges downward 4".

11. Do the same for the other side.

12. Extrude faces to the outside by 5".

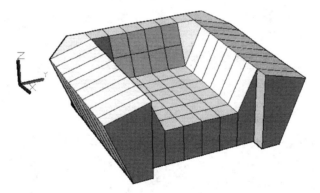

13. Merge the faces, as shown below, and move it downward 3".

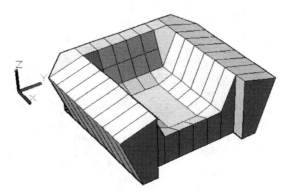

14. Move the faces upward 6", as shown below:

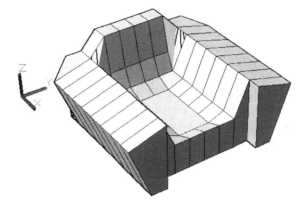

15. Change the smoothness of the shape to Level 4.

16. Select the faces, as shown below, and set the crease value to 1.

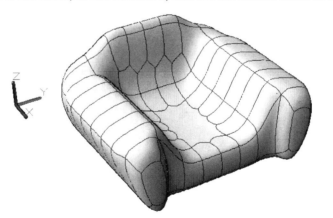

17. Save and close the file.

CHAPTER REVIEW

1. SURFTAB1 and SURFTAB2 will affect:

    a. Tabulated surface and Ruled Surface

    b. Tabulated surface, Revolved surface, and Ruled surface

    c. Tabulated surface, Revolved surface, Ruled surface, and Edge surface

    d. None of the above

2. There are three levels of smoothness for meshes.

   a. True

   b. False

3. Faces, edges, and vertices can be controlled by you, but you cannot control _____ in meshes.

4. In order to see the internal edges of a mesh, which system variable should you change?

   a. VSEDGECOLOR

   b. VSEDGELEX

   c. VSEDGES

   d. EDGEDISPLAY

5. In a Revolved surface, the axis of revolution could be any two points.

   a. True

   b. False

6. _____ command adds sharpness to the mesh.

7. Ruled surfaces can deal with:

   a. Closed 2D objects

   b. Points

   c. Open 2D objects

   d. All of the above

8. You can delete a face of a mesh by:

   a. Using the DELMESHFACE command

   b. Selecting the face and pressing [Del] on the keyboard

   c. Going to the Mesh tab, locating the Mesh Edit panel, and clicking the Delete button

   d. You cannot delete a single mesh face

CHAPTER REVIEW ANSWERS

1. c

3. Facets

5. b

7. d

NOTES

9. You should have created something similar to the following:

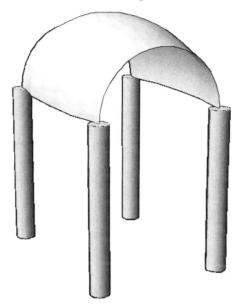

10. Save and close the file.

4.3 BLEND, PATCH, AND OFFSET COMMANDS

Blend, Patch, and Offset commands enable the creation of complex surfaces.

4.3.1 Blend Command

This command creates a continuity surface between existing surface edges. To issue this command, go to the **Surface** tab, locate the **Create** panel, and select the **Blend** button:

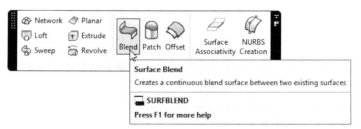

The following prompts are displayed:

```
Continuity = G1 - tangent, bulge magnitude = 0.5
Select first surface edges to blend or [Chain]:
Select first surface edges to blend or [Chain]:
Select second surface edges to blend or [Chain]:
Select second surface edges to blend or [Chain]:
```

The first line presents the current values for **Continuity** and **bulge magnitude**. AutoCAD asks you to select the edges in two sets; once you have selected the first set of edges, press [Enter] and select the second set of edges. When done, press [Enter]. Use the **Chain** option to select all adjacent edges. When finished, the following prompts are displayed:

```
Press Enter to accept the blend surface or
[CONtinuity/Bulge magnitude]:
```

Press [Enter] to accept the default values, or change the values to something else. You will see the following:

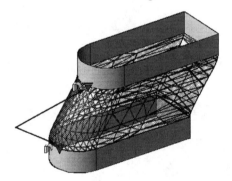

The two triangles allow you to change the value of continuity at the two edges selected:

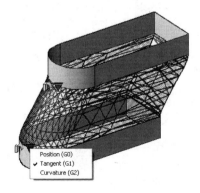

Continuity values are:

* G0 – Position

* G1 – Tangent

* G2 – Curvature

G0 is a flat surface. G1 creates a new surface that is tangent to the two edges selected. Finally, G2 creates the best curvature. You can change the Bulge Magnitude value to manage the amount of swelling. The default value is 0.5; the lowest value is 0 and the highest value is 1. This is the final result:

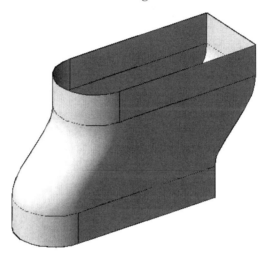

Continuity = G2, Bulge = 0.7

4.3.2 Patch Command

This command creates a surface that serves as a "cover" (from the top or bottom) for an existing surface. To issue this command, go to the **Surface** tab, locate the **Create** panel, and select the **Patch** button:

The following prompts are displayed:

```
Continuity = G0 - position, bulge magnitude = 0.5
Select surface edges to patch or [CHain/CUrves]
<CUrves>:
```

The first line presents the current values for continuity and bulge magnitude. AutoCAD then asks you to select the surface edges to be patched. When done, press [Enter]. Use the Chain option to select all adjacent edges at the same time. The following prompts are shown:

```
Press Enter to accept the patch surface or
[CONtinuity/Bulge magnitude/Guides]:
```

Press [Enter] to accept the default values, or change the values to new values. The other option is **Guides**, for which the following prompt is shown:

```
Select curves or points to constrain patch
surface:
```

Select the object for the patching process. See the following illustrations. This first example is without Guides:

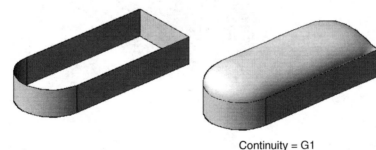

Continuity = G1

This example is with a guide:

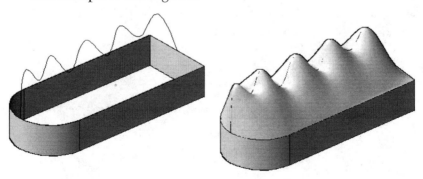

4.3.3 Offset Command

This command creates a parallel copy of an existing surface. To issue this command, go to the **Surface** tab, locate the **Create** panel, and select the **Offset** button:

The following prompts are shown:

```
Connect adjacent edges = No
Select surfaces or regions to offset:
```

The first line presents the current value for Connect adjacent edges. AutoCAD then asks you to select the desired surface(s). When done, press [Enter].

By selecting a surface, you will see something similar to the following picture:

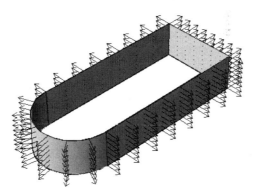

The following prompts are shown:

```
Specify offset distance or [Flip direction/Both
sides /Solid/Connect] <0.0000>:
```

The picture above shows the default offset direction, but you also have the following options:

- Flip direction flips the direction of the offset.
- Both sides specifies the offset to be set to both sides.
- Solid creates a new solid object from the outcome of the offset.
- Connect connects multiple offset surfaces if the original surfaces were connected.

PRACTICE 4-2

Blend, Patch, and Offset Commands

1. Start AutoCAD 2015.

2. Open Practice 4-2.dwg.

3. Offset the upper surface to the outside by distance = 3.

4. Offset the lower surface to the inside by distance = 3, keeping the Connect option = Yes.

5. Blend the inner surfaces (using Chain in the lower surface), keeping Continuity = G1 and Bulge = 0.5 for both edges.

6. Blend the outer surfaces (using Chain in the lower surface), making Continuity = G2 for both edges.

7. Patch the lower two shapes using Continuity = G0.

8. Thaw the Curve layer.

9. Using the Patch command and Guide option, patch the inner surface using the curve, without changing Continuity and Bulge.

10. Freeze the Curve layer.

11. Using the Blend command, blend the edges of the two surfaces without changing Continuity and Bulge.

Before Untrim   After Untrim

4.4.4 Extend Command

This command lengthens the selected surface's edges. To issue this command, go to the **Surface** tab, locate the **Edit** tab, and select the **Extend** button:

The following prompts are shown:

```
Modes = Extend,  Creation = Append
Select surface edges to extend:
Select surface edges to extend:
Specify extend distance or [Modes]:
```

The first line presents the current values for mode and creation. AutoCAD then asks you to select the desired edge(s). When done, press [Enter]. As a final step, specify the extend distance.

You can control the Extension mode as follows:

• Extend mode increases the length of the surface and attempts to replicate and continue the current surface shape.

• Stretch mode increases the length of the surface but does not try to replicate and continue the current surface shape.

You can control the Creation type as follows:

- Merge merges surfaces into a single surface.
- Append maintains both new and old surfaces.

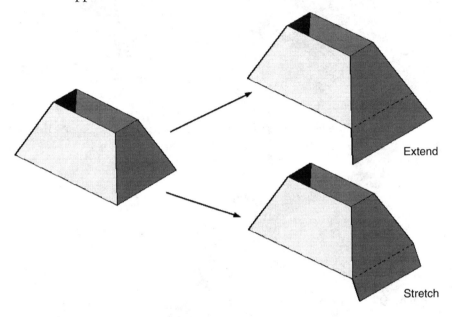

Extend

Stretch

4.4.5 Sculpt Command

This command converts a surface into a solid. To issue this command, go to the **Surface** tab, locate the **Edit** tab, and select the **Sculpt** button:

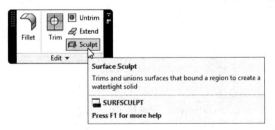

The following prompts are shown:

```
Mesh conversion set to: Smooth and optimized.
Select surfaces or solids to sculpt into a solid:
```

A **watertight** volume is the condition for a successful conversion process.

4.4.6 Editing Surfaces Using Grips

You can adjust the value of the fillet radius after the command has finished, but only if the surface created by fillet is **not** NURBS. To do this, simply click the surface and the fillet icon is displayed for further editing. See the following illustration:

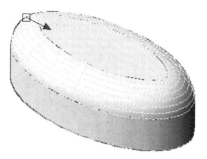

Additionally, you can use grips to edit a surface created from an extrusion (discussed further in the next chapter). You can manipulate both the taper angle and height. See the following illustrations:

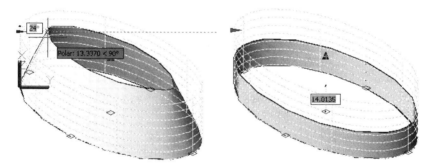

You can change the blend, patch, and offset values. See the following example:

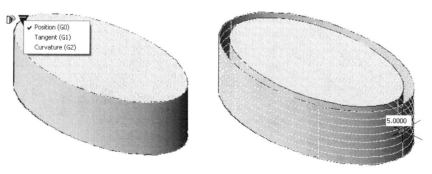

The Quick Properties and Properties palettes provide information about how the surface was created, and if it is an extrusion, offset, Associative, or NURBS. See the following examples of the Quick Properties and Properties palettes that display the origin of the surface:

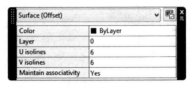

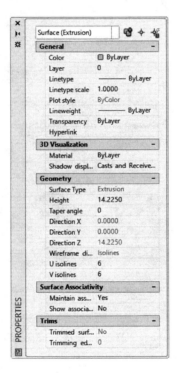

In Properties, you can change the Height, Taper angle, whether or not Associativity is maintained, and whether or not the surface is trimmed.

PRACTICE 4-3

Editing Surfaces

1. Start AutoCAD 2015.

2. Open Practice 4-3.dwg.

3. Make sure Surface Associativity is on.

4. Select the entire shape.

5. Using the arrow on the left, change the taper angle to 20°.

6. Increase the height to 25 units.

7. Using the Extend command, extend the four lower edges (each one in a separate command) by 15 units with the Stretch and Merge options.

8. Using the Blend command, blend the edges between the extended surfaces.

9. Thaw the 2D Shapes layer, and trim using the polygon in the two directions.

10. Try to change one of the two shapes and note how the trimmed shape changes. Undo what you did.

11. Try to erase one of the two shapes and note how the trimming disappears. Undo what you did.

12. Freeze the 2D Shapes layer.

13. Using the Patch command, make covers on the top and bottom using the default values.

14. Try to convert the surface to a solid using the Sculpt command. The operation will not work due to the openings.

15. Untrim the openings.

16. Using the Sculpt command again, try to convert it into solid. Did it work?

17. The final shape should look similar to the following:

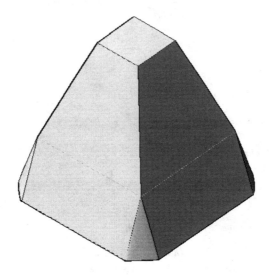

18. Save and close the file.

PRACTICE 4-4

Fillet and Grips Editing

1. Start AutoCAD 2015.

2. Open Practice 4-4.dwg.

3. Using the Fillet command in the Surface tab, connect the two surfaces at the top and middle using radius = 2, and make sure that Trim Surface = Yes.

4. Do the same to fillet the other two surfaces, but this time with radius = 3.

5. Change the radius of first bend to 3 using the graphical method.

6. Your final shape should be similar the following:

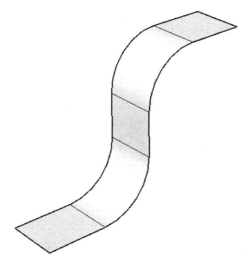

7. Save and close the file.

4.5 PROJECT GEOMETRY

AutoCAD enables the projection of 2D shapes on 3D surfaces and solids. If desired, the projected shape can trim the surfaces and solids. All commands are found in the **Surface** tab, on the **Project Geometry** panel, as shown here:

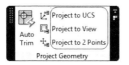

As you can see, the big button titled Auto Trim is an on/off button. If on, the projection trims the surface or solid. The three commands are:

- Project to UCS
- Project to View
- Project to 2 Points

4.5.1 Project to UCS

This option projects the 2D shape along the +ve/-ve Z-axis of the current UCS. The following prompts are displayed:

```
Select curves, points to be projected or
[PROjection direction]:
Select a solid, surface, or region for the target
of the projection
```

The first prompt asks you to select the desired 2D shape. You can select multi objects. When done, press [Enter]. The second prompt asks you to select the solid or surface (you can select a 2D region as well). See the following illustration (notice the current UCS):

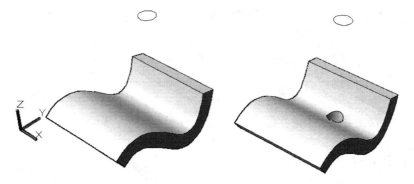

4.5.2 Project to View

This option projects the geometry based on the current view. The following prompts are displayed:

```
Select curves, points to be projected or
[PROjection direction]:
Select a solid, surface, or region for the target
of the projection:
```

These are the same as the first option's prompts. See the following illustration and note that the trimming process is not affected by the current UCS:

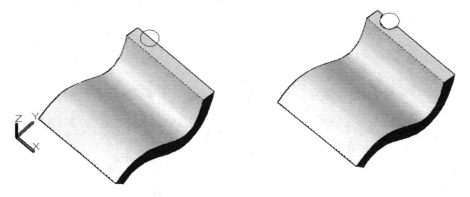

The following prompt is displayed:

```
Select an NURBS surface or curve to edit:
Select point on NURBS surface.
```

The first prompt asks you to select the surface. Then specify a point on the NURBS surface for the new vertex location. The result is something similar to the following:

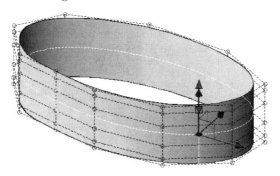

Click the small triangle and the following is displayed:

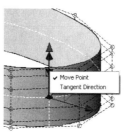

You can do three things:

- Move the new vertex:

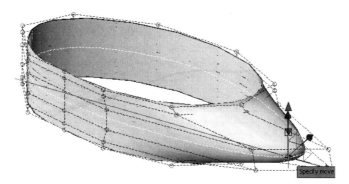

- Change the tangency:

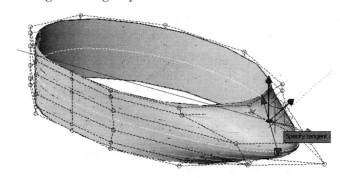

- Use the arrow to edit the tangent magnitude:

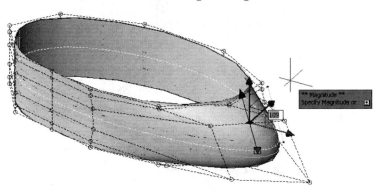

As the final step, after you have finished creating your surfaces, analyze them. To do this, turn Hardware acceleration on using the status bar:

Be aware that turning this button on reduces the speed of your computer. Now, go to the **Surface** tab. There is a panel called **Analysis,** which has four buttons:

There are three analysis methods by which you can analyze your surface:

- Zebra Analysis tests surface continuity. See the following example. Since the continuity is G0, the two surfaces are not tangent. This is evident because the lines are not aligned between the two surfaces:

- Curvature Analysis provides Gaussian, minimum, maximum, and mean values for surface curvature. Maximum curvature and positive Gaussian values are displayed in green, while minimum curvature and negative Gaussian values are displayed in blue. Planes, cylinders, and cones have no Gaussian curvature.

- Draft Analysis displays a color gradient on a surface to assess if there is sufficient space between a part and its mold.

To control these analyses, select the **Analysis Options** button. The following dialog box is displayed:

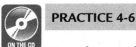

PRACTICE 4-6

Creating and Editing NURBS

1. Start AutoCAD 2015.

2. Open Practice 4-6.dwg.

3. This is a Procedural surface.

4. Convert it to an NURBS surface.

5. Show CV.

6. Try to move one of the CVs. Does AutoCAD allow you to do this?

7. Rebuild the surface as degree 3, with the number of vertices in U = 6 and in V = 10.

8. Remove the two sets of CVs, as indicated below (this is top view):

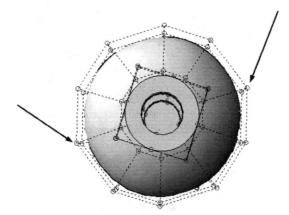

9. Go to one of the two sides from which you removed the CV, and, using the CV Edit bar, add a new CV near the center of the rectangle formed by the four CVs. Change the magnitude to be 175.

10. Do the same for the other side.

11. Patch the bottom of the shape.

12. Hide CVs.

13. The shape will look like the following:

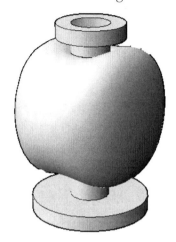

14. Save and close the file.

NOTES

To issue this command, go to the **Solid** tab, locate the **Solid** panel, and select the **Extrude** button; or go to the **Surface** tab, locate the **Create** panel, and select the **Extrude** button:

The following prompts are shown:

```
Select objects to extrude or [MOde]:
Specify height of extrusion or [Direction/Path/
Taper angle]
```

The first line asks you to select the object(s) to extrude. The second line says:

```
Specify height of extrusion or [Direction/Path/
Taper angle]
```

This line indicates there are four ways to create an extrusion in AutoCAD. They are:

- Specifying a height
- Specifying a direction
- Specifying a path
- Specifying a taper angle

5.4.1 Create an Extrusion Using Height

This is the default option, and it creates an extrusion perpendicular to the 2D profile using height. If you select the 2D object, edge, or face, the extrusion is displayed on the screen. You can specify the height graphically or by typing in a value. See the following:

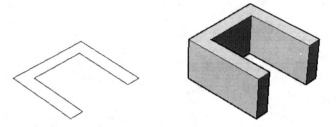

5.4.2 Create an Extrusion Using a Direction

The default height option creates an extrusion perpendicular to the profile, but this option creates an extrusion with any angle desired. The following prompts are shown:

```
Specify start point of direction:
Specify end point of direction:
```

The following picture illustrates the concept:

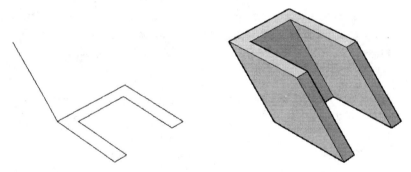

5.4.3 Create an Extrusion Using a Path

This option creates an extrusion using a path. You can use almost all 2D objects as paths, as well as a 3D Poly, Helix, and the edges of a surface or solid. Before selecting the path you can set the taper angle. The following prompts are shown:

```
Select extrusion path or [Taper angle]:
```

Selecting the taper angle option displays the following prompt:

```
Specify angle of taper for extrusion or
[Expression]:
```

The following picture illustrates the selection of a path:

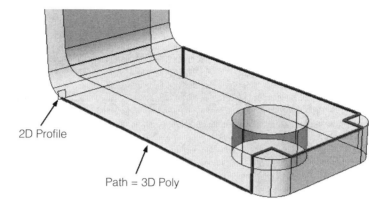

2D Profile

Path = 3D Poly

This results in something similar to the following:

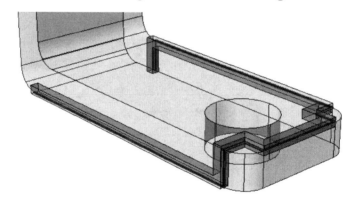

This example illustrates the use of a path with a taper angle:

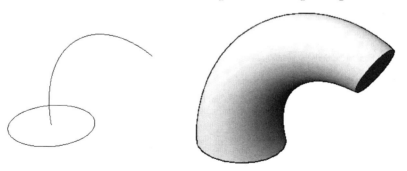

Taper angle = 5

5.4.4 Create an Extrusion Using a Taper Angle

This option sets a positive taper angle (to the inside) or negative taper angle (to the outside). Once the taper angle is set the same prompts are displayed again, meaning the taper angle can work with all of the above options (height, direction, and path). See the following prompts:

```
Specify angle of taper for extrusion or [Expression]:
Specify height of extrusion or [Direction/Path/Taper
angle/Expression]
```

Taper angle = 15

PRACTICE 5-2

ON THE CD

Extrude Command

1. Start AutoCAD 2015.

2. Open Practice 5-2.dwg.

3. Change the visual style to Conceptual.

4. Extrude the two rectangles upward, with height = 200 and taper angle = 1.

5. Thaw the Path layer.

6. Start the Extrude command and select the top face of one of the pillars (don't use Subobjects = Face, as this prevents you from selecting the path); change the taper angle = 0.

7. Union the three solids.

8. You will get the following shape:

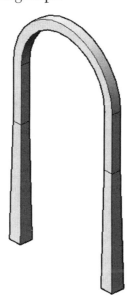

9. Save and close the file.

5.5 LOFT COMMAND

This command creates a solid or surface from multiple cross sections of different UCSs. All profiles must be opened or closed; you cannot use a mix of opened and closed. You can use different options such as guide, path, and cross sections. To issue this command, go to the **Solid** tab, locate the **Solid** panel, and select the **Loft** button; or go to the **Surface** tab, locate the **Create** panel, and select the **Loft** button:

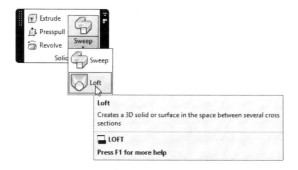

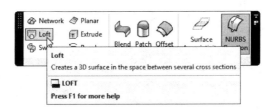

The following prompts are displayed:

```
Select cross sections in lofting order or [POint/
Join multiple edges/MOde]:
```

The first line asks you to select the cross sections in lofting order. See the following example:

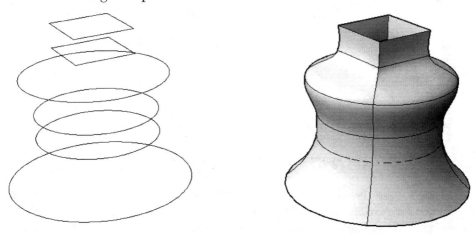

This prompt is repeated so you can select more closed or opened objects. When selecting, you have three options, explained here:

- The point option selects the loft end point and ends the command at this point. The cross section selected should always be closed, and you should specify this point by typing a real 3D point or by selecting a point object. The following prompt is displayed:

```
Specify loft end point:
```

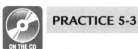

PRACTICE 5-3

Loft Command

1. Start AutoCAD 2015.

2. Open Practice 5-3.dwg.

3. Change the visual style to Conceptual.

4. Go to the Surface tab.

5. Using the Loft command, loft the three circles with the path that represents the handle of the ancient jug.

6. Using the Loft command, select the circles one by one from top to bottom. When done, press [Enter]. Select Normal for all sections and press [Enter] to execute the command.

7. To clean the part of the handle inside the jug, go to the Surface tab, locate the Edit panel, and click the Trim command.

8. Now select the handle and press [Enter].

9. Select the jug and press [Enter].

10. When asked to Select an area to trim, zoom to the lower part of the handle and click beside it.

11. Patch the jug from the bottom using Continuity = G0.

12. You should have created something similar to the following:

13. Save and close the file.

5.6 REVOLVE COMMAND

This command creates a solid or surface by revolving open or closed 2D objects, or by using the face or edge of solids and surfaces. To issue this command, go to the **Solid** tab, locate the **Solid** panel, and select the **Revolve** button; or go to the **Surface** tab, locate the **Create** panel, and select the **Revolve** button:

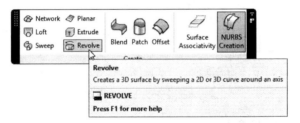

The following prompt is displayed:

```
Select objects to revolve or [MOde]:
```

AutoCAD asks you to select the desired object(s) to revolve. When done, press [Enter]. The following prompts are displayed:

```
Specify axis start point or define axis by
[Object/X/Y/Z] <Object>:
Specify angle of revolution or [STart angle/
Reverse/EXpression] <360>:
```

AutoCAD offers three methods to specify the axis of revolution. They are:

- Specify an axis by inputting two points

- Specify an axis by selecting a drawn object

- Specify an axis by specifying one of the three axes

Finally, specify the angle of revolution. There are multiple options:

- Input the angle by typing or by using the mouse. CCW is always positive. If you define the axis of revolution by two points, the positive angle is represented by the curl of the fingers of your right hand when your thumb is pointing from the direction of the nearest point to the farthest point.

- Specify the start angle if you want the revolving process to start somewhere other than from the cross section.

- Use the Reverse option to reverse the direction of the revolution.

The following illustration explains the concept:

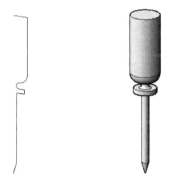

PRACTICE 5-4

Revolve Command

1. Start AutoCAD 2015.

2. Open Practice 5-4.dwg.

3. Revolve the shape on the left using two points as the axis of revolution, picking from top to bottom, and assigning 180° as the rotation angle.

4. Undo the last step.

5. Revolve the shape on the left using two points as the axis of revolution, picking from bottom to top, and assigning 180° as the rotation angle. Did you see a difference? Why did this happen?

6. Revolve the shape on the right using a line as an axis of revolution with 360° as the rotation angle.

7. The final result should look similar to the following:

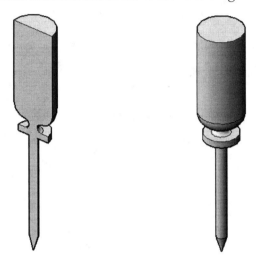

8. Save and close the file.

5.7 SWEEP COMMAND

This command creates complex solids or surfaces by sweeping an open or closed 2D shape, solid, surface face, or edge along a path. To issue this command, go to the **Solid** tab, locate the **Solid** panel, and select the **Sweep** button; or go to the **Surface** tab, locate the **Create** panel, and select the **Sweep** button:

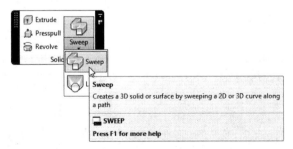

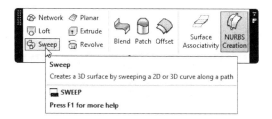

The following prompts are shown:

```
Select objects to sweep or [MOde]:
Select objects to sweep or [MOde]:
Select sweep path or [Alignment/Base point/
Scale /Twist]:
```

The default option selects an object or objects to sweep, and then selects the sweep path. If you follow the path below, the following is the result:

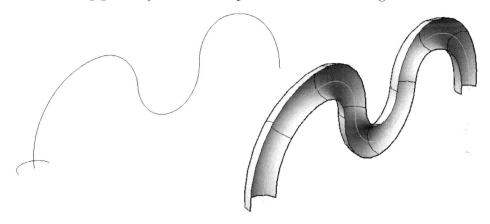

Other options include:

- Alignment
- Base point
- Scale
- Twist

5.7.1 Alignment Option

This option sets the relationships between the profile and the path. Sometimes the path is normal to the profile, and sometimes it is not! See the following two examples. The first example illustrates a path that is normal to the profile:

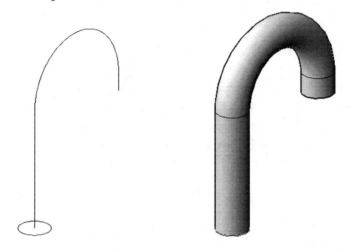

This example illustrates a path that is not normal to the profile:

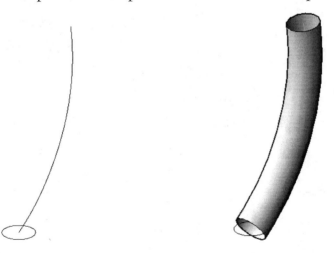

In the second example, if you turn Alignment off you will get the following result:

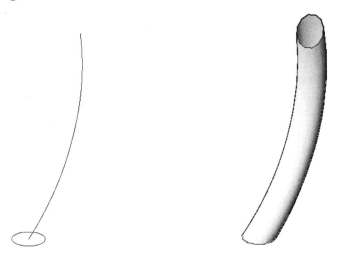

The following prompt is displayed:

```
Align sweep object perpendicular to path before
sweep [Yes/No]<Yes>:
```

5.7.2 Base Point Option

This option is used to specify a new base point for the solid or surface that differs from the location of the profile object. The following prompt is displayed:

```
Specify base point:
```

Specify the coordinates by typing them in or by using the mouse. The following example illustrates this concept:

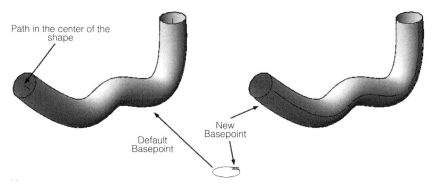

5.7.3 Scale Option

This option sets a scale factor for the end of the sweeping solid or surface. The following prompt is displayed:

```
Enter scale factor or [Reference/
Expression]<1.0000>:
```

Input the desired scale and press [Enter]. The result is something similar to the following:

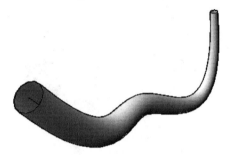

5.7.4 Twist Option

This option sets an angle of twist while sweeping. The following prompt is displayed:

```
Enter twist angle or allow banking for a non-
planar sweep path [Bank/EXpression]<0.0000>:
```

Input the desired twist angle. The result is something similar the following:

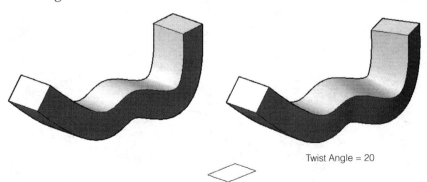

Twist Angle = 20

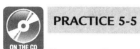

PRACTICE 5-5

Sweep Command

1. Start AutoCAD 2015.

2. Open Practice 5-5.dwg.

3. Thaw the Helix layer and 2D Object layer. Turn on Show/Hide Lineweight.

4. Zoom to the groove at the bottom.

5. Using the Sweep command, use the circle as a profile and the helix as the path, without changing the defaults.

6. Subtract the new shape from the solid.

7. Repeat this with the top groove but use Scale = 0.5.

8. Freeze the Helix layer.

9. Zoom to the lower left corner of the top shape to see the rectangle there.

10. Using the Sweep command (from the Surface tab), use the rectangle as a profile and the edge of the solid as the path (you need to change the Subobjects filter to Edge).

11. Repeat this for the other side.

12. The final shape should look similar to the following:

13. Save and close the file.

5.8 EDITING SOLIDS AND SURFACES USING THE FOUR COMMANDS

Once you have created a solid or surface, clicking the object provides three editing methods:

- Using Grips
- Using Quick Properties
- Using Properties

5.8.1 Using Grips

Clicking a solid or surface created from an extrusion displays the following:

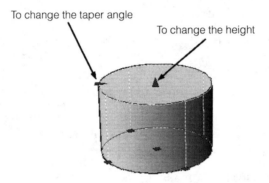

Clicking a solid or surface created from loft displays the following:

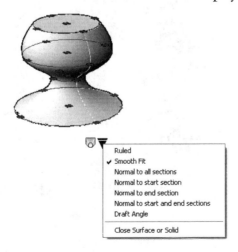

Clicking a solid or surface created by the Revolve command displays the following:

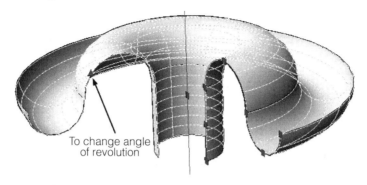

To change angle of revolution

Clicking a solid or surface created from sweep displays the following:

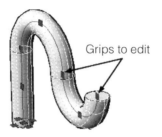

Grips to edit

5.8.2 Using Quick Properties

Quick Properties display different types of information for solids and surfaces. The following two examples apply to surfaces:

Surface (Extrusion)	
Color	☐ ByLayer
Layer	0
Height	16.5254
Taper angle	0
U isolines	6
V isolines	6
Maintain associativity	Yes

Surface (Sweep)	
Color	☐ ByLayer
Layer	0
Profile rotation	0
Scale along path	1.0000
Twist along path	0
U isolines	6
Maintain associativity	Yes

And the following apply to solids:

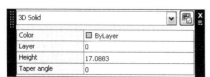

3D Solid	
Color	☐ ByLayer
Layer	0
Height	17.0883
Taper angle	0

3D Solid	
Color	☐ ByLayer
Layer	0
Profile rotation	0
Twist along path	0
Scale along path	1.0000

5.8.3 Using Properties

The Properties palette displays the following:

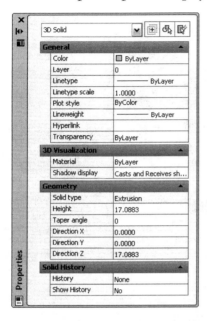
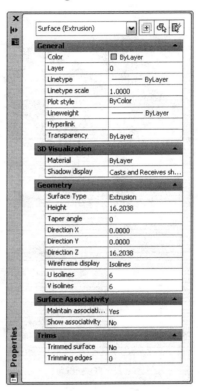

For both Quick Properties and Properties, there is more information presented for surfaces than for solids.

5.9 DELOBJ SYSTEM VARIABLE

Using 2D objects with Extrude, Loft, Revolve, and Sweep produces a solid or surface, but what happens to the 2D object? Does it remain? Is it erased? Variable DELOBJ answers this question. It controls whether the 2D object is deleted or not. If DELOBJ = 0, the profiles, cross sections, paths, and Guide curves remain. If DELOBJ = 1, the profiles alone are deleted. If DELOBJ = 2, the profiles, paths, and Guide curves are deleted.

PROJECT 5-1-M

Project Using Metric Units

1. Start AutoCAD 2015.

2. Open Project 5-1-M.dwg.

3. For each part of the drawing there is a layer in the same name. Make sure you are in the right layer when you draw the shape.

4. Draw a box 30 m x 30 m x 0.9 m.

5. At the center of the top face of the box, draw a cylinder R = 3 m and Height = 6 m.

6. Using one of the quadrants of the top face of the cylinder, draw a rectangle 0.15 m x 0.15 m. If drawn outside, move it to the inside using the midpoint of one of the sides, so it coincides with the quadrant of the cylinder.

7. Using the Surface tab, extrude the rectangle by 0.6 m.

8. Using the Polar Array, create an array from the small columns with 30° angles between each shape.

9. You will have the following shape:

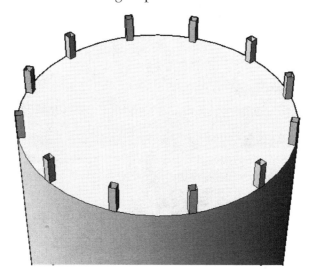

10. Draw a line between the top midpoints of any two columns.

11. Using Gizmo, move it down 0.3 m.

12. Draw the line again.

13. Draw a circle R = 0.01 m at the end of one of the lines. Do the same for the other line.

14. Using the Surface tab, sweep the circle using the line as your path. Repeat the same for the other line.

15. Using the Polar array and the same input you used for the columns, array the protecting ropes.

16. The shape should look like the following:

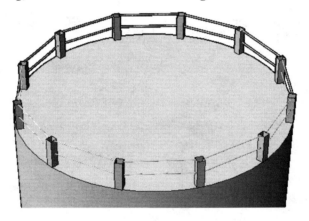

17. Draw a circle at the center of the top face of the cylinder R = 0.6 m.

18. Using Presspull, pull it upward 0.12 m.

19. At the top of the new shape, draw another circle R = 0.3 m.

20. Change the UCS to look like the following:

21. Draw the following polyline:

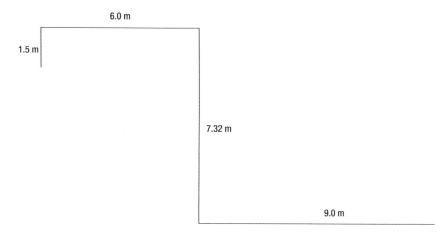

6.0 m

1.5 m

7.32 m

9.0 m

22. Using the Fillet command, fillet the polyline using Radius = 0.5 m, and use the Polyline option to do it in one shot.

23. Using the Surface tab, extrude the circle with R = 0.3 m, using the polyline as your path.

24. The final shape should look like the following:

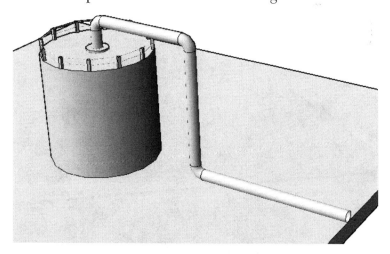

25. Save and close the file.

PROJECT 5-1-I

Project Using Imperial Units

1. Start AutoCAD 2015.

2. Open Project 5-1-I.dwg.

3. For each part of the drawing there is a layer of the same name. Make sure you are in the right layer when you draw the shape.

4. Draw a box 100' x 100' x 3'.

5. At the center of the top face of the box, draw a cylinder R = 10' and Height = 20'.

6. Using one of the quadrants of the top face of the cylinder, draw a rectangle 6" x 6". If drawn outside, move it to the inside using the midpoint of one of the sides, so it coincides with the quadrant of the cylinder.

7. Using the Surface tab, extrude the rectangle by 2'.

8. Using the Polar Array, create an array from the small columns with a 30° angle between each shape.

9. You will have the following shape:

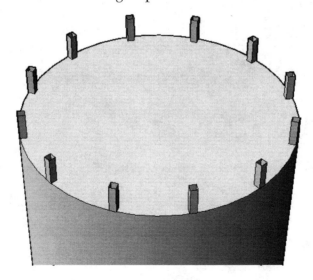

10. Draw a line between the top midpoints of any two columns.

11. Using gizmo, move it down 1'.

12. Draw the line again.

13. Draw a circle R = 0.4" at the end of one of the lines. Do the same to the other line.

14. Using the Surface tab, sweep the circle using the line as your path. Repeat the same for the other line.

15. Using the Polar array and the same input you used for the columns, array the protecting ropes.

16. The shape should look like the following:

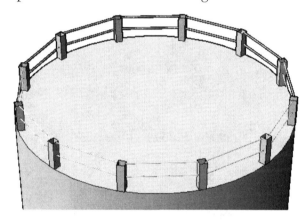

17. Draw a circle at the center of the top face of the cylinder R = 2'.

18. Using Presspull, pull it upward 5".

19. At the top of the new shape, draw another circle R = 1'.

20. Change the UCS to look like the following:

21. Draw the following polyline:

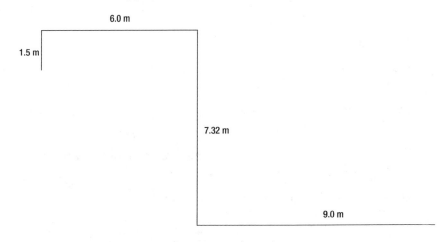

22. Using the Fillet command, fillet the polyline using Radius = 20", and the option Polyline to do it in one shot.

23. Using the Surface tab, extrude the circle with R = 1', using the polyline as your path.

24. The final shape should look like the following:

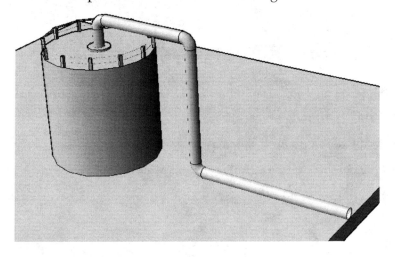

25. Save and close the file.

NOTES

CHAPTER REVIEW

1. Which one of the following statements is false?

    a. DELOBJ has three different values.

    b. Quick Properties and Properties display different information for Surfaces and Solids.

    c. You can't consider line as a sweeping path.

    d. You can delete the profile after finishing the revolution using the DELOBJ system variable.

2. Twist and scale are options in the _____ command.

3. Which one of the following is not an option in the Loft command?

    a. Path

    b. Guide

    c. Normal to all sections

    d. Normal to no section

4. While using the Revolve command, the axis of revolution should always be a drawn object.

    a. True

    b. False

5. You can change the taper angle when using the grips of an extrusion.

    a. True

    b. False

6. The default twist of a helix is _____.

CHAPTER REVIEW ANSWERS

1. c

3. d

5. a

CHAPTER 6

SOLID EDITING COMMANDS

In This Chapter

- Solid Face Manipulation Commands
- Solid Edge Manipulation Commands
- Solid Body Manipulation Commands

6.1 INTRODUCTION

There is a specific command in AutoCAD that enables the editing of everything related to a solid object: edges, faces, and the entire body of the solid. This command is **solidedit**, and is accessed using the command window and keyboard. This command is long and tedious. To make your life easier, AutoCAD customized all **solidedit** options in the ribbon interface. Go to the **Home** tab, locate the **Solid Editing** panel, and you can find the commands.

Face commands:

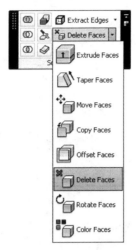

Edge commands:

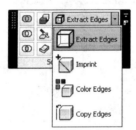

Body commands:

These commands are discussed in the following pages.

6.2 EXTRUDING SOLID FACES

This command extrudes the selected *planar* faces of a solid. You can use different methods to extrude a face, such as using the height, path, or taper angle. If you input a positive value, AutoCAD increases the volume of the solid. A negative value indicates a decrease in the volume of the solid. (The reverse increases an opening in a solid).

To issue this command, go to the **Home** tab, locate the **Solid Editing** panel, and select the **Extrude Faces** button:

The following prompts are displayed:

```
Select faces or [Undo/Remove]:
Select faces or [Undo/Remove/ALL]:
Specify height of extrusion or [Path]:
Specify angle of taper for extrusion <0>:
Solid validation started.
Solid validation completed.
```

The best way to select a solid face is to hold [Ctrl] when selecting it or when changing a subobject to Face. The remaining prompts have been explained in the discussion on the Extrude command. Press [Enter] twice to end this command. See the following:

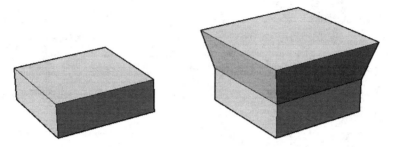

6.3 TAPERING SOLID FACES

This command creates an inclined face using two points. A positive angle means the face will incline inward, and a negative angle means it will incline outward. To issue this command, go to the **Home** tab, locate the **Solid Editing** panel, and select the **Taper Faces** button:

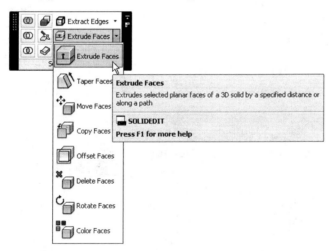

The following prompts are shown:

```
Select faces or [Undo/Remove]:
Select faces or [Undo/Remove/ALL]:
Specify the base point:
Specify another point along the axis of tapering:
Specify the taper angle:
Solid validation started.
Solid validation completed.
```

Select Faces in the filter pull down in the Subobject panel. To make your selection process easy, hold the [Ctrl] key as you select the faces. After selecting the faces, specify a base point and a second point to define the axis of tapering, input the tapering angle, and press [Enter] twice. See the following example (this is a positive angle, so the face is inclined inward):

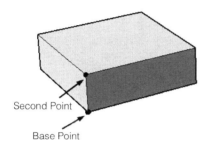
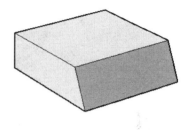

PRACTICE 6-1

Extruding and Tapering Solid Faces

1. Start AutoCAD 2015.

2. Open Practice 6-1.dwg.

3. Zoom to the solid on the left.

4. Extrude the top face with height = 240 and taper angle = +5.

5. Do the same for the solid on the right.

6. Thaw the Path layer.

7. Extrude the top face of either shape using the path shown (use any method to select; to select Subobject, select No Filter before selecting the path).

8. Using the Taper Face command, taper the three front faces of the two bases (the shapes from the practice) using a negative angle = -15 (it should be to the outside).

9. The final result should look like the following:

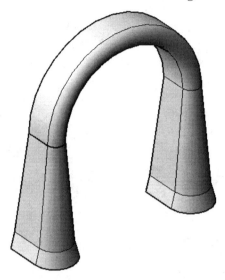

10. Save and close the file.

6.4 MOVING SOLID FACES

This command uses a base point to move faces of a solid. To issue this command, go to the **Home** tab, locate the **Solid Editing** panel, and select the **Move Faces** button:

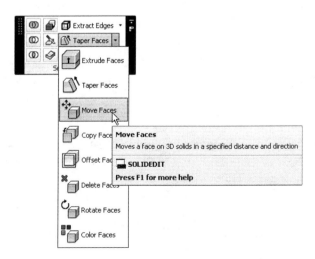

The following prompts are displayed:

```
Select faces or [Undo/Remove]:
Select faces or [Undo/Remove/ALL]:
Specify a base point or displacement:
Specify a second point of displacement:
Solid validation started.
Solid validation completed.
```

Select faces by using the Face option in the Subobject panel or by holding [Ctrl] while making your selection. Specify a base point, and then specify a second point (same as moving in 2D). Finally, press [Enter] twice to execute the command.

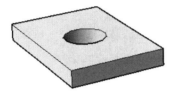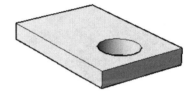

6.5 COPYING SOLID FACES

This command copies the faces of a solid by using a base point to produce a region or surface. To issue this command, go to the **Home** tab, locate the **Solid Editing** panel, and select the **Copy Faces** button:

The following prompts are displayed:

```
Select faces or [Undo/Remove/ALL]:
Select faces or [Undo/Remove/ALL]:
Specify a base point or displacement:
Specify a second point of displacement:
Enter a face editing option
```

Select faces by using the Face option in the Subobject panel or by holding [Ctrl] while making your selection. Specify a base point, and then specify a second point (same as copying in 2D). Finally, press [Enter] twice to execute the command. This results is a **region** object for planar faces and a **surface** for curved faces.

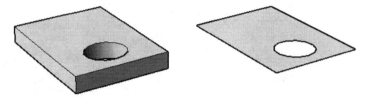

PRACTICE 6-2

Moving and Copying Solid Faces

1. Start AutoCAD 2015.

2. Open Practice 6-2.dwg.

3. Move the hole from its center to the center of the curved shape.

4. Copy the following two faces:

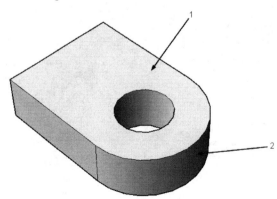

5. What is the object type of the copied face from (1)? _____
(Region)

6. What is the object type of the copied face from (2)? _____
(Surface)

7. Save and close the file.

6.6 OFFSETTING SOLID FACES

This command offsets solid faces. The offset command differs from the extrude command in that the offset command can offset both planar and curved faces. Input a positive offset distance and the volume increases, and input a negative offset distance and the volume decreases. To issue this command, go to the **Home** tab, locate the **Solid Editing** panel, and select the **Offset Faces** button:

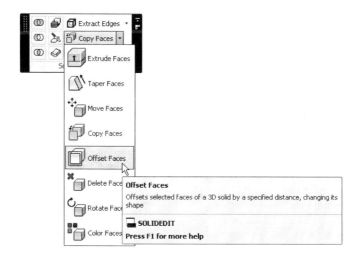

The following prompts are displayed:

```
Select faces or [Undo/Remove]:
Select faces or [Undo/Remove/ALL]:
Specify the offset distance:
Solid validation started.
Solid validation completed.
```

Select faces using the Face option in the Subobject panel or by holding [Ctrl] while making your selection. Then specify the offset distance. Finally, press [Enter] twice to execute the command.

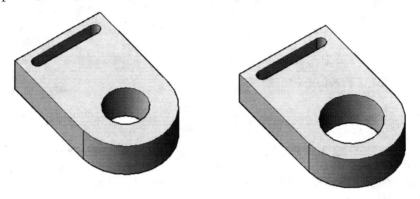

6.7 DELETING SOLID FACES

This command deletes a solid face. If you delete a face and AutoCAD is unable to close its gap, the process terminates and the following prompt is displayed:

```
Modeling Operation Error:
Gap cannot be filled.
```

To issue this command, go to the **Home** tab, locate the **Solid Editing** panel, and select the **Delete Faces** button:

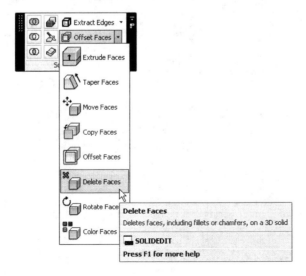

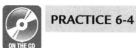

PRACTICE 6-4

Rotating and Coloring Solid Faces

1. Start AutoCAD 2015.

2. Open Practice 6-4.dwg.

3. Using Presspull, pull the right part of the top face of the cylinder upward 6 units.

4. Rotate the two faces by 15° to produce the shape as follows:

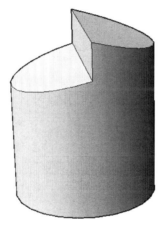

5. Color the two faces red.

6. Save and close the file.

6.10 EXTRACTING SOLID EDGES

This command creates a wireframe of lines from the edges of a selected solid. To issue this command, go to the **Home** tab, locate the **Solid Editing** panel, and select the **Extract Edges** button:

The following prompts are displayed:

```
Select objects:
Select objects:
```

AutoCAD asks you to select the solids. When done, press [Enter]. The output is in the same location as the original solid, so use the Move command to move the solid so that the extracted edges are displayed. The following picture illustrates this concept:

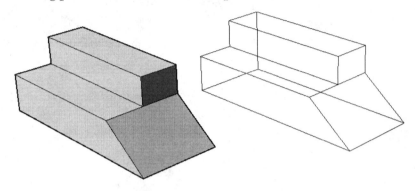

6.11 EXTRACT ISOLINES

This command extracts an isoline curve from a surface, a solid, or a face of a solid. To issue this command, go to the **Surface** tab, locate the **Curves** panel, and select the **Extract Isolines** button:

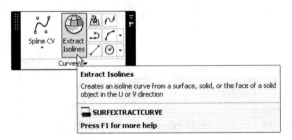

The following prompt is displayed:

```
Select a surface, solid, or face:
Extracting isoline curve in U direction
Select point on surface or [Chain/Direction/Spline
points]:
```

The first line asks you to select a surface, a solid, or a face of a solid. Once this is done, AutoCAD reports the direction in which the curve is will be extracted: U or V (in the above example, the curve is extracted in U direction). Finally, AutoCAD asks you to specify a point on the surface to finalize the extraction.

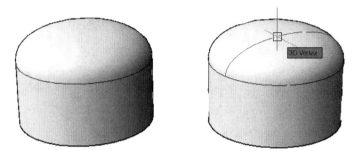

There are three more available options: Chain, which creates a chain of isolines; Direction, to change the direction of extraction from U to V and vice-versa; and Spline points, which specifies points on the curved surface on which to draw a spline.

Isolines that are extracted from Surfaces are called splines (they look like circles), and isolines that are extracted from Solids are lines, arcs, and circles.

6.12 IMPRINTING SOLIDS

This command cuts an existing face into more faces by using 2D objects as imprints. To issue this command, go to the **Home** tab, locate the **Solid Editing** panel, and select the **Imprint** button:

The following prompts are displayed:

```
Select a 3D solid or surface:
Select an object to imprint:
Delete the source object [Yes/No] <N>:
```

This command affects both surfaces and solids, which makes it unique. A 2D object should be drawn prior to using this command. AutoCAD asks you to select the solid or surface, and then the 2D object. Finally, specify whether or not to keep the 2D object. The following picture illustrates this concept:

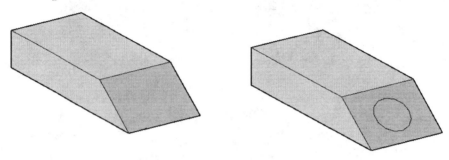

6.13 COLORING SOLID EDGES

This command assigns colors to selected edges. To issue this command, go to the **Home** tab, locate the **Solid Editing** panel, and select the **Color Edges** button:

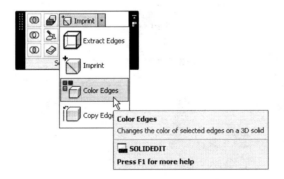

The following prompts are shown:

```
Select edges or [Undo/Remove]:
Select edges or [Undo/Remove/ALL]:
```

Select the edges to be colored. When done, press [Enter]. The Select Color dialog box is shown. Select the desired color, and click OK. The colored edges are shown in the following visual styles: 2D Wireframe, Hidden, Shaded with edges, and Wireframe.

The following picture illustrates this concept:

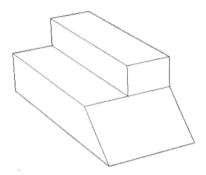

6.14 COPYING SOLID EDGES

This command copies the edges of a solid to produce a 2D object. It is similar to the Extract edges command discussed above, but it differs in that it gives you the freedom to select some of the edges instead of all of them. To issue this command, go to the **Home** tab, locate the **Solid Editing** panel, and select the **Copy Edges** button:

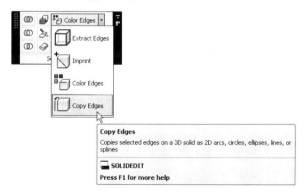

The following prompts are shown:

```
Select edges or [Undo/Remove]:
Specify a base point or displacement:
Specify a second point of displacement:
```

AutoCAD asks you to select the edges, and a base point and second point that are identical to that in the new copy of the object. When done, press [Enter] twice. The following picture explains this concept:

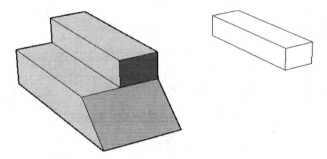

6.15 OFFSETTING SOLID EDGES

This command offsets a planar solid edge to create a closed polyline or spline. To issue this command, go to the **Solid** tab, locate **Solid Editing**, and select the **Offset Edge** button:

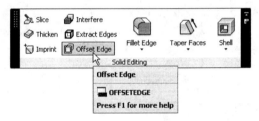

The following prompts are shown:

```
Select face:
Specify through point or [Distance/Corner]:
```

AutoCAD asks you to select a face and then a through point. This is the default option. Specify the distance and side of offset. Another method is to select the Distance option, which displays the following two prompts:

```
Specify distance <0.0000>:
Specify point on side to offset:
```

Input the desired distance and specify the side to offset. The final option is Corner, which specifies the type of corners that appear on the offset object if it extends beyond the selected face. The choices are Sharp or Rounded. The output of this command can be used in the Presspull Extrude command. The following picture illustrates this concept:

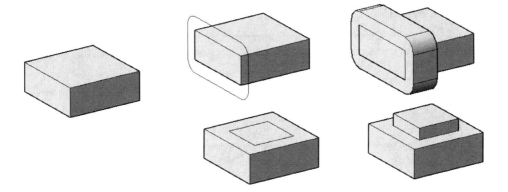

PRACTICE 6-5

Solid Edge Functions

1. Start AutoCAD 2015.

2. Open Practice 6-5.dwg.

3. Using the small rectangle, imprint the big face without keeping the 2D object.

4. Using Extrude Faces, Extrude the newly created face using the polyline as a path. If you find it hard to select the small face to extrude, try selecting the small face by selecting one of its edges. This method selects both faces. Then select the Remove option and remove the big face from the selection set.

5. This is what you should have:

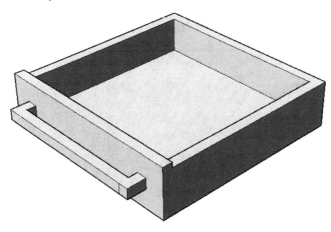

6. Color all the edges red.

7. Change the visual style to Hidden.

8. Using Extracting Edges, extract all edges of the drawer and move the resultant lines to the right in a shape similar to >.

9. Change the visual style to Conceptual.

10. Thaw the Surface layer.

11. Using the Extract Isolines command, extract isolines in U direction and V direction. Move the surface so you can see the isolines on their own. What type of object do the isolines form? _____

12. Save and close the file.

6.16 SEPARATING SOLIDS

Use of the subtract command can sometimes result in two or more non-continuous solids. This command fixes this problem by separating these portions so that each one forms an independent single solid. To issue this command, go to the **Home** tab, locate the **Solid Editing** panel, and select the **Separate** button:

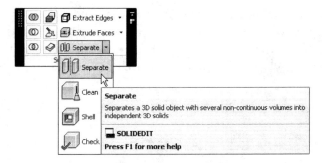

The following prompts are shown:

```
Select a 3D solid:
```

AutoCAD asks you to select the solids to be separated. When done, press [Enter] twice to execute the command. The following picture illustrates this concept:

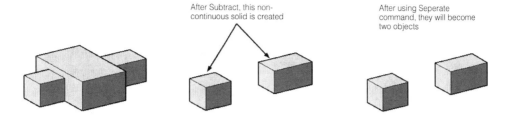

After Subtract, this non-continuous solid is created

After using Seperate command, they will become two objects

6.17 SHELLING SOLIDS

This command hollows a solid. To issue this command, go to the **Home** tab, locate the **Solid Editing** panel, and select the **Shell** button:

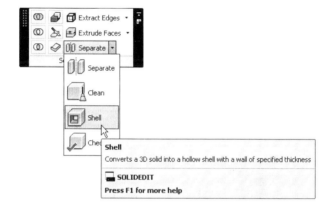

The following prompts are shown:

```
Select a 3D solid:
Remove faces or [Undo/Add/ALL]:
Remove faces or [Undo/Add/ALL]:
Enter the shell offset distance:
Solid validation started.
Solid validation completed.
```

AutoCAD asks you to select the desired solid. You can remove faces from the object to see within it, as though it were a hollowed out shell. AutoCAD then asks you to input the offset distance. Inputting a positive value means the outside edge of the solid remains outside, and the offset distance will be to the inside. Inputting a negative value means the outside edge of the solid will be the inside edge and the offset will be to the outside. See the following illustration:

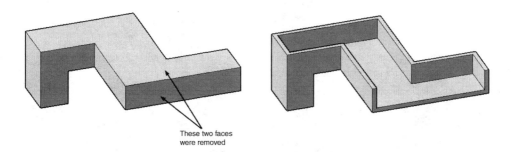

These two faces
were removed

There are two other commands for solids:

- Clean removes any redundant edges/vertices on a solid.

- Check tells you if the selected objects are solids or another 3D object.

PRACTICE 6-6

Separating and Shelling Solids

1. Start AutoCAD 2015.

2. Open Practice 6-6.dwg.

3. Using the Shell command, shell the pipe to the inside with Offset distance = 1.

4. The pipe should look like the following:

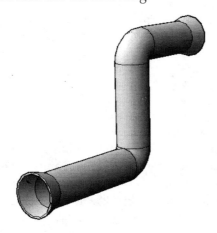

SOLID EDITING COMMANDS • 203

6. Use the _____ command when there are two or more non-continuous solid volumes.

7. With the Extrude face command, you can use Path with Taper angle:

    a. True

    b. False

8. Sometimes AutoCAD refuses to delete a selected face.

    a. True

    b. False

CHAPTER REVIEW ANSWERS

    1. b

    3. a

    5. b

    7. b

NOTES

CHAPTER 7

3D MODIFYING COMMANDS

In This Chapter

- 3D Move, 3D Rotate, and 3D Scale
- 3D Array and 3D Align
- Filleting and Chamfering Solids

7.1 3D MOVE, 3D ROTATE, 3D SCALE, SUBOBJECTS, AND GIZMO

Older versions of AutoCAD used 3D Move, 3D Rotate, and 3D Scale. Subobjects and Gizmo now perform the same functions as these three commands, and are covered in detail in Chapter 2 and Chapter 3. This section covers the older 3D Move, 3D Rotate, and 3D Scale commands to provide a foundation of information, but it is highly recommended that you use Subobjects and Gizmo instead.

7.1.1 3D Move Command

To issue this command, go to the **Home** tab, locate the **Modify** panel, and select the **3D Move** button:

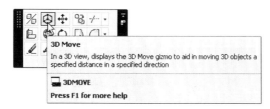

The following prompts are shown:

```
Select objects:
Specify base point or [Displacement] <Displacement>:
Specify second point or <use first point as
displacement>:
```

Select the desired objects or subobjects and perform the move.

7.1.2 3D Rotate Command

To issue this command, go to the **Home** tab, locate the **Modify** panel, and select the **3D Rotate** button:

The following prompts are shown:

```
Select objects:
Specify base point:
Pick a rotation axis:
Specify angle start point or type an angle:
```

Select the desired objects or subobjects and perform the rotation.

7.1.3 3D Scale Command

To issue this command, go to the **Home** tab, locate the **Modify** panel, and select the **3D Scale** button:

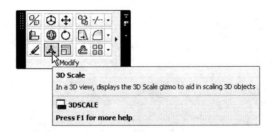

The following prompts are shown:

```
Select objects:
Specify base point:
Pick a scale axis or plane:
Specify scale factor or [Copy/Reference] <1.0000>:
```

Select the desired objects or subobjects and perform the scaling.

7.2 MIRRORING IN 3D

This command uses a mirror plane to create a mirror image in the 3D space. This differs from the normal Mirror command, which only works in the current XY plane. To issue this command, go to the **Home** tab, locate the **Modify** panel, and select the **3D Mirror** button:

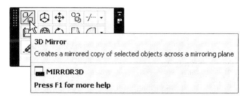

The following prompts are shown:

```
Select objects:
Select objects:
Specify first point of mirror plane (3 points) or
[Object/Last/Zaxis/View/XY/YZ/ZX/3points] <3points>:
```

AutoCAD offers multiple ways to create a mirror plane, some of which are the same as the UCS options.

7.2.1 XY, YZ, and ZX Options

These three options are very similar and are therefore presented here together. They enable the creation of a mirroring plane that is parallel to the original three planes. See the following prompts:

```
[Object/Last/Zaxis/View/XY/YZ/ZX/3points] <3points>:
Specify point on ZX plane <0,0,0>:
Delete source objects? [Yes/No] <N>:
```

The ZX option is used in the above example. AutoCAD asks you to specify a point on the selected plane. Specify your point.

7.2.2 3 Points Option

You can create a plane by specifying 3 points (this option is the same as the 3 points option in the UCS command). This option is the most generic, as it gives you the freedom to create any mirroring plane. The following prompts are shown:

```
[Object/Last/Zaxis/View/XY/YZ/ZX/3points]
<3points>:
Specify first point on mirror plane:
Specify second point on mirror plane:
Specify third point on mirror plane:
Delete source objects? [Yes/No] <N>:
```

7.2.3 Last Option

This option retrieves the last used mirroring plane.

7.2.4 Object, Z-axis, View options

These options were discussed in the section on UCS commands.

For all of these options, the last prompt asks whether or not to delete the source object.

PRACTICE 7-1

ON THE CD

Mirroring in 3D

1. Start AutoCAD 2015.

2. Open Practice 7-1.dwg.

3. Using the 3D Mirror command, mirror the shape using different methods (minimum two).

4. Union the two shapes.

5. You should get the following shape:

6. Save and close the file.

7.3 ARRAYING IN 3D

The same command to array in 2D is used to array in 3D, but with some modifications. There are three different arraying options in AutoCAD:

- Rectangular Array
- Path Array
- Polar Array

These options are associative, so you are able to edit the values of the array after the command is finished. To issue this command, go to the **Home** tab, locate the **Modify** panel, and select one of the three commands:

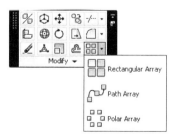

Once you have finished a normal 2D associative array of any type (rectangular, path, or polar), click it, and a new context tab named **Array** is displayed:

This context tab provides two areas that can be used to make your rectangular array act as a 3D array. The first is the **Levels** panel, in which you can enter a third dimension for the array. Input the following:

- Level Count

- Level Spacing

- Total Level Distance

The second area is the **Rows** panel. Extend the panel and the following is displayed:

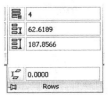

Based on the elevation of the first row, **Incremental Elevation** assigns an increasing elevation the second row, and then third row, and so on. The two pictures below illustrate the difference between levels and incremental elevation (the example used here is a rectangular array):

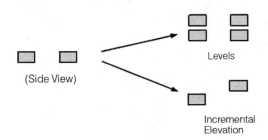

7.5 FILLETING SOLID EDGES

This command fillets a solid edge. It is possible to use the 2D Fillet command, but it is recommended that you use this command instead. To issue this command, go to the **Solid** tab, locate the **Solid Editing** panel, and select the **Fillet Edge** button:

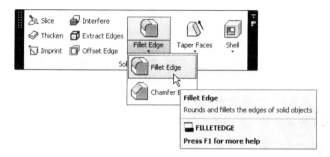

The following prompts are shown:

```
Radius = 1.0000
Select an edge or [Chain/Loop/Radius]:
Select an edge or [Chain/Loop/Radius]:
2 edge(s) selected for fillet.
```

The first line reports the current value of the radius. AutoCAD then asks you to select the edges. If you don't change the radius value it uses the default value. Once you have selected the edges, press [Enter]. AutoCAD reports the number of edges selected. The following message is shown:

```
Press Enter to accept the fillet or [Radius]:
```

Press [Enter] to accept the fillet produced, or change the **Radius** value by moving the radius value arrow or by selecting the **Radius** option and typing in a new value.

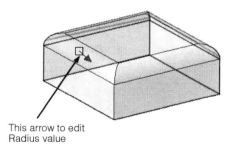

This arrow to edit
Radius value

7.5.1 Chain Option

Use this option to select more than one edge if the edges are tangent to one another. Select this option and the following prompt is displayed:

```
Select an edge chain or [Edge/Radius]:
```

The condition for success is tangency. See the following illustration of the Chain concept:

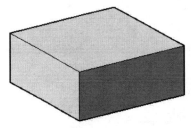

Edges are not tangent to each

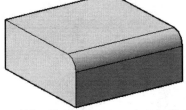

Using Chain option, one edge was filleted

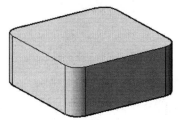

Edges are tangent to each other

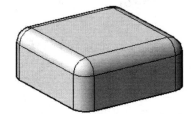

Using Chain option, all edges were filleted

7.5.2 Loop Option

This option selects a loop of edges whether they are tangent to each other or not. There are two possible loops for any given edge. AutoCAD randomly selects one of these two loops, and asks you to accept it or select the next loop. The following illustration shows the first step after an edge is selected:

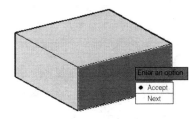

Select the Next option and all edges are filleted at once:

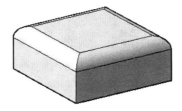

7.6 CHAMFERING SOLID EDGES

This command uses distances to chamfer selected solid edges. To issue this command, go to the **Solid** tab, locate the **Solid Editing** panel, and select the **Chamfer Edge** button:

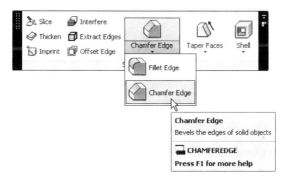

The following prompts are shown:

```
Distance1 = 1.0000, Distance2 = 1.0000
Select an edge or [Loop/Distance]:
Select an edge belongs to the same face or [Loop/
Distance]:
Select an edge belongs to the same face or [Loop/
Distance]:
```

The first line reports the current values of the two distances. AutoCAD then asks you to select the edges. If you don't change the distance values the default values are used. When you are done selecting edges, press [Enter]. The following message is shown:

```
Press Enter to accept the chamfer or [Distance]:
```

Press [Enter] to accept the chamfer produced, or change the **Distance** values by moving the two chamfer distance value arrows or by selecting the **Distance** option and typing in two new values. The following picture illustrates this concept:

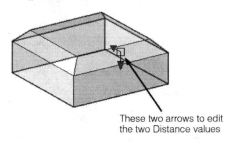

These two arrows to edit
the two Distance values

The loop option is the same as it is for the Fillet Edge command.

PRACTICE 7-3

Aligning, Filleting, and Chamfering

1. Start AutoCAD 2015.

2. Open Practice 7-3.dwg.

3. Using Align 3D, align the Cylinder to the center of the box, as follows:

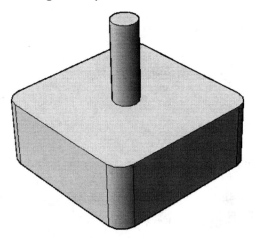

4. Look at the model using the Top view and Parallel view.

5. Move the cylinder 16 units to the right, then 16 units up.

6. Copy the cylinder to the four edges of the box using 32 units in all directions.

7. Using Gizmo and the Move command, move the four cylinders down 25 units.

8. Subtract the four cylinders from the big shape.

9. Using the Fillet Edge command and the Chain option, fillet the upper and lower edges using radius = 2.5 units.

10. Using the Chamfer Edge command, and Distance 1 = Distance 2 = 2 units, chamfer the top edges of the four cylinders.

11. The final shape should look similar to the following:

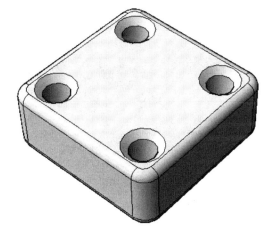

12. Save and close the file.

PROJECT 7-1-M

Project: Using Metric Units

1. Start AutoCAD 2015.

2. Open the file Project 7-1-M.dwg.

3. Using Taper Faces, taper the two faces by angle = 15° to look like the following (the lower edges should go toward the outside):

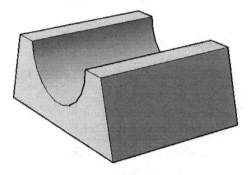

4. Thaw the Path layer.

5. Extrude the front face using the path, and you should get the following shape:

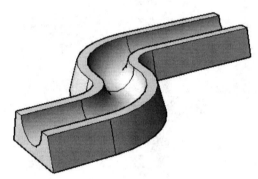

6. Using Mirror 3D, mirror the shape using the ZX plane as the mirroring plane.

7. Union the two shapes.

8. Thaw the Circle and Path layer.

9. Using the Extrude command, extrude the circle using the path.

10. Move the pipes to the center of the base.

11. Shell the pipes to the inside by 6 units.

12. You should get the following shape:

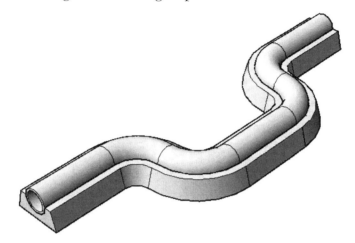

13. Using the Quadrant Osnap, draw a line, as shown:

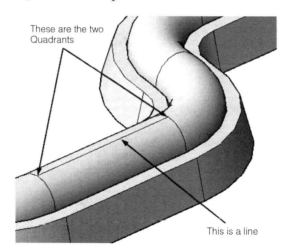

These are the two Quadrants

This is a line

14. Using the two ends and the midpoint of the line, draw three circles with radius = 32 units.

15. Erase the line.

16. Using the Presspull command, press the three circles to the inside by 32 units, then subtract the new shapes from the pipe.

17. Using the Extrude command, extrude the two circles at the two ends upward by 260 units.

18. It appears that the hole in the middle is unnecessary, so delete the face along with the circle.

19. Using Gizmo, move the two cylinders downward 20 units.

20. Using Offset Edge, offset the two top faces of the two newly created cylinders to the inside using distance = 6 units.

21. Using Presspull, press the two circles by 260 units to shell the two cylinders.

22. The shape should look like the following:

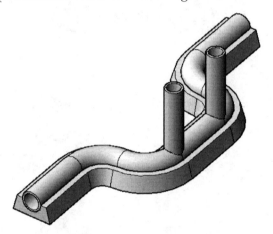

23. Draw a line from the edge, as shown below, in the positive X direction (WCS) of distance = 400 unit.

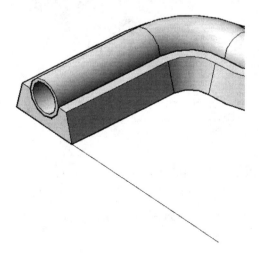

24. Using the Polyline command, draw the following shape at the edge of the line (all the lines are 8 units, except the long ones they are 24 units).

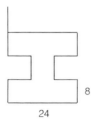

25. Using the Extrude command, extrude the polyline up 480 units.

26. Using 3D Array (Rectangular), create an array with the following information:

    a. Number of rows = 4

    b. Number of columns = 2

    c. Number of levels = 2

    d. Row spacing = 400

    e. Column spacing = -800

    f. Level spacing = 510

27. Draw a box 824 x 1224 x 30 and copy it to cover the first set of columns, and then copy it again to cover the second set of columns.

28. Fillet the edge as shown below using R = 12 units.

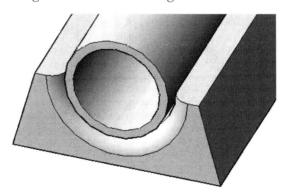

29. Do the same thing for the other end.

30. Using Fillet edge command and R = 6 and Chain option, fillet the edge as in the following picture:

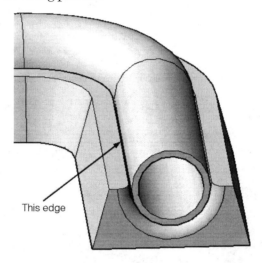

This edge

31. Do the same for the other edges.

32. The final shape should look like the following:

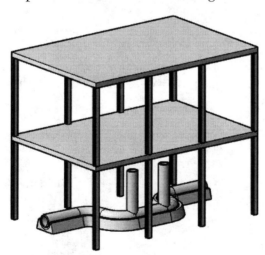

33. Save and close the file.

PROJECT 7-1-I

Project Using Imperial Units

1. Start AutoCAD 2015.

2. Open file Project 7-1-I.dwg.

3. Using Taper Faces, taper the two faces by angle = 15° to look like the following (the lower edges should go to the outside):

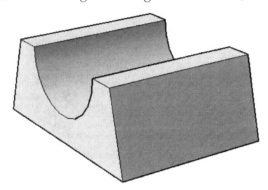

4. Thaw the Path layer.

5. Extrude the front face using the path. You should get the following shape:

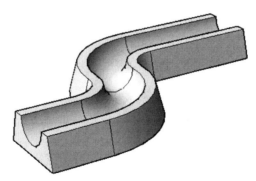

6. Using Mirror 3D, mirror the shape using ZX plane as the mirroring plane.

7. Union the two shapes.

8. Thaw the Circle and Path layers.

9. Using the Extrude command, extrude the circle using the path.

10. Move the pipes to the center of the base.

11. Shell the pipes to the inside by 3".

12. You should get the following shape:

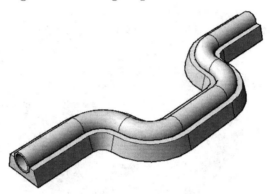

13. Using the Quadrant Osnap, draw a line as shown:

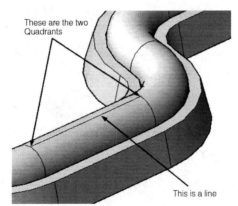

These are the two Quadrants

This is a line

14. Using the two ends and midpoint of the line, draw three circles R = 13".

15. Erase the line.

16. Using the Presspull command, press the three circles to the inside by 13", then subtract the new shapes from the pipe.

17. Using the Extrude command, extrude the two circles at the two ends up by 8'8".

18. As for the hole in the middle, it appears that we don't need it, so Delete the face along with the circle.

19. Using Gizmo, move the two cylinders downward by 8".

20. Using Offset Edge, offset the two top faces of the two newly created cylinders to the inside using distance = 3".

21. Using Presspull, press the two circles by 8'8" to shell the two cylinders.

22. The shape should look like the following:

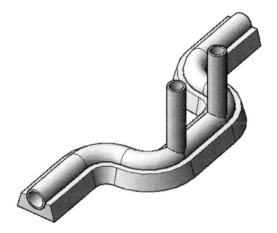

23. Draw a line from the edge, as shown below, in the positive X direction (WCS) of distance = 14'.

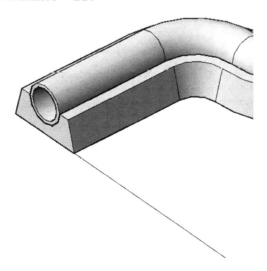

24. Using the Polyline command, draw the following shape at the edge of the line (all the lines are 3" except the long ones, which are 10").

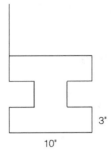

10"

3"

25. Using the Extrude command, extrude the polyline up 16'.

26. Using 3D Array (Rectangular), create an array with the following information:

    a. Number of rows = 4

    b. Number of columns = 2

    c. Number of levels = 2

    d. Row spacing = 14'

    e. Column spacing = -28'

    f. Level spacing = 17'

27. Draw a box 28'9" x 42'10" x 1'0", copy it to cover the first set of columns, and then copy it again to cover the second set of columns.

28. Fillet the edge, as shown below, using R = 5".

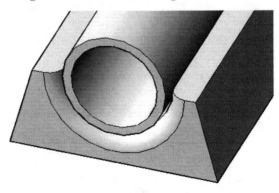

29. Do the same for the other end.

30. Using the Fillet edge command and R = 3", and the Chain option, fillet the edge, as in the following picture:

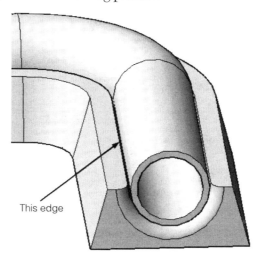

This edge

31. Do the same for the other edges.

32. The final shape should look like the following:

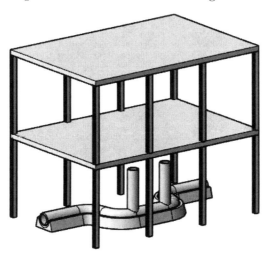

33. Save and close the file.

NOTES

CHAPTER REVIEW

1. You can delete a fillet from a solid.

    a. True

    b. False

2. There are special commands to fillet and chamfer solid edges.

    a. True

    b. True, but only for tangent edges.

    c. True, but you can still use the normal Fillet and Chamfer commands as in 2D.

    d. False

3. Using the Array command in a 3D environment, you can set _____ and _____ to utilize the 3D feature of this command.

4. When using the 3D Mirror command:

    a. Specify a mirror plane using 3 points.

    b. Specify a mirror line just as in 2D.

    c. Specify a mirror plane using XY, YZ, ZX, or any plane that is parallel to them.

    d. A & C

5. For the Chain option in the Fillet Edge command to be successful, the edges should be tangent to each other.

    a. True

    b. False

6. For the Loop option in the Chamfer Edge command to be successful, the edges should be tangent to each other:

    a. True

    b. False

CHAPTER REVIEW ANSWERS

1. a

3. Levels, Incremental Elevation

5. a

CHAPTER 8

CONVERTING AND SECTIONING

In This Chapter

- Converting Objects
- Sectioning using Slice, Section Plane, and Flatshot Commands

8.1 INTRODUCTION: CONVERTING OBJECTS

Previous chapters have outlined the different types of 3D objects in AutoCAD. The discussion covered some techniques to convert some 2D and 3D objects to solids, meshes, and surfaces. The techniques discussed include:

- Converting 2D objects to solids or surfaces using Extrude, Loft, Revolve, and Sweep commands

- Converting 2D objects to solids the Using Presspull command

- Converting solids to meshes using the Smooth command

- Converting solids, meshes, and surfaces to an NURBS Surface using the Convert to NURBS command

- Converting surfaces to solids using the Sculpt command

This is actually only a small number of the available conversion techniques. This section presents the remaining techniques, giving you the information you need to use the tools to solve any problem.

8.2 CONVERTING 2D AND 3D OBJECTS TO SOLIDS

This section discusses the Thicken command, which converts any surface to a solid. The second command, Convert to Solid, is a generic command that can convert meshes and some 2D objects to solids.

Here are the two commands:

8.2.1 Thicken Command

This command converts any surface to a solid. To issue this command, go to the **Home** tab, locate the **Solid Editing** panel, and select the **Thicken** button:

The following prompts are shown:

```
Select surfaces to thicken:
Select surfaces to thicken:
Specify thickness:
```

AutoCAD asks you to select the surface to thicken. You can select any number of surfaces. When done, press [Enter], and input the value of the thickness. The following picture illustrates this concept:

8.4 CONCLUSION: CONVERTING OBJECTS

In conclusion, see the following table of commands that can be used to convert any object to another:

	From 2D objects	*From Mesh*	*From Solid*	*From Surface*
To Solid	Extrude, Loft, Revolve, Sweep, Presspull	Convert to Solid	—	Sculpt, Thicken, Convert to Solid
To Surface	Extrude, Loft, Revolve, Sweep,	Convert to Surface, Convert to NURBS	Convert to Surface, Convert to NURBS, Explode (some objects only)	—
To Mesh	—	—	Smooth	Smooth

PRACTICE 8-1

ON THE CD

Converting Objects

1. Start AutoCAD 2015.

2. Open Practice 8-1.dwg.

3. There is a mesh sphere. Go to the Mesh tab, locate the Convert Mesh panel, select Smoothed Not Optimized, and select the Convert to Solid command.

4. Undo what you did.

5. Select Smoothed Optimized and then select the Convert to Solid command.

6. Compare the number of faces in both cases. Which has more faces? _____ Why? _____

7. Freeze the Mesh layer and Thaw the Surface layer.

8. Set VSEDGES to 1.

9. Use the Thicken command with thickness = 5.

10. Change the visual style to X-Ray. Do you see a hollow sphere with thickness?

11. Change the visual style to Conceptual.

12. Freeze the Surface layer and Thaw the Solid layer.

13. Explode the solid.

14. What is the resultant object? _____

15. Save and close the file.

8.5 SECTIONING IN 3D

This section presents three commands that can produce a section of a 3D model. They are:

- Slice: Works on solids and surfaces, but not on meshes

- Section Plane: Works on all types of 3D objects

- Flatshot: Creates a view parallel to the current view of any 3D object

Each command is unique, and can help create 2D sections of 3D shapes with their different methods.

8.6 SECTIONING USING THE SLICE COMMAND

This command uses a slicing plane to create a 3D slice of a 3D object. To issue this command, go to the **Home** tab, locate the **Solid Editing** panel, and select the **Slice** button:

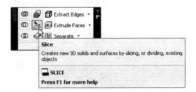

The following prompts are shown:

```
Select objects to slice:
Specify start point of slicing plane or [planar
Object/Surface/Zaxis/View/XY/YZ/ZX/3points]
```

```
<3points>:
Specify a point on the ZX-plane <0,0,0>:
Specify a point on desired side or [keep Both
sides] <Both>:
```

AutoCAD asks you to select the solid or surface, and specify a slicing plane. The Surface option aligns the slicing plane to a surface (Revsurf, Tabsurf, Rulesurf, or Edgesurf). The other methods are discussed in the section on the Mirror 3D command in Chapter 7. Lastly, AutoCAD asks you to specify the side on which you wish to stay; or you can choose to keep both sides.

PRACTICE 8-2

Slicing Objects

1. Start AutoCAD 2015.

2. Open Practice 8-2.dwg.

3. Use the Slice command to achieve the following result:

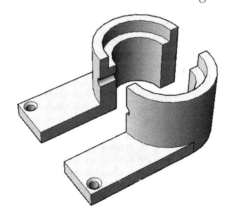

4. Save and close the file.

8.7 SECTIONING USING SECTION PLANE COMMAND

This command creates a 2D or 3D section from a 3D object. It has the ability to insert the section as a block, or export it to another file. To issue this command, go to the **Home** tab, locate the **Section** panel, and select the **Section Plane** button:

The following prompts are shown:

```
Select face or any point to locate section line
or [Draw section/ Orthographic]:
```

Three methods can create a section plane. They are:

8.7.1 Select a Face

This is the default option. You are asked to select one of the faces as the section plane. The following picture illustrates this concept:

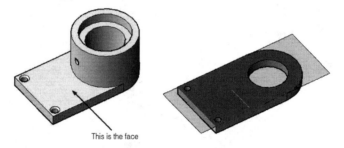

This is the face

8.7.2 Draw Section

This option provides the ability to draw a section plane using Osnap. The following prompts are shown:

```
Specify start point:
Specify next point:
Specify next point or ENTER to complete:
Specify next point or ENTER to complete:
Specify point in direction of section view:
```

AutoCAD asks you to specify points as you would when drafting a line or polyline. When done, press [Enter]. AutoCAD needs a minimum of two points to draw a section plane. Then click the side you want the sectioning to work on. The following picture illustrates this concept:

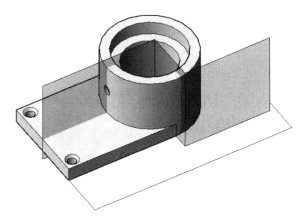

8.7.3 Orthographic

This option creates a section plane parallel to one of the six orthographic planes relative to the current UCS. The following prompt is shown:

```
Align section to: [Front/bAck/Top/Bottom/Left/
Right]:
```

Select the desired plane. The section plane is implemented on the 3D object.

The following picture explains this concept:

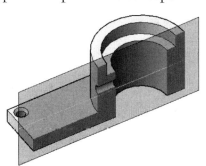

This is aligned with Right view of the WCS

8.7.4 Section Options

After a section plane has been created, click the plane to view the three available options. The triangle shown below is displayed; clicking it displays the following:

Three section options are displayed:

- Section Plane (default option) displays only the section plane. See the following illustration:

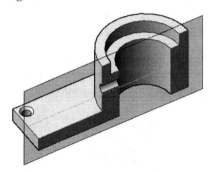

- Section Boundary sets the 2D range to control what is affected by the section plane. See the following illustration:

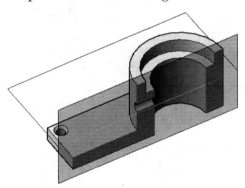

- Section Volume sets the 3D range to control which objects are affected by the section plane. See the following illustration:

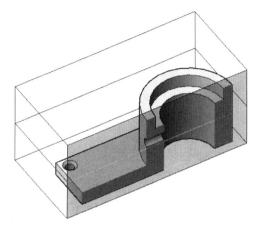

8.7.5 Section Plane Grips

After you have created and controlled the section plane, you can manipulate it further using grips. Click on the section plane and something similar to the following is displayed:

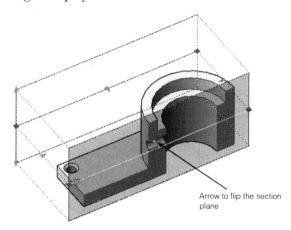

Arrow to flip the section plane

Use the rectangle and triangle grips to control the volume of the section plane. Use the arrow to flip the side of the section.

8.7.6 Adding a Jog to an Existing Section Plane

This command adds a jog to an existing section plane (you can still draw a jog using the Draw section option discussed above). To issue this command, go to the **Home** tab, locate the **Section** panel, and select the **Add Jog** button:

The following prompts are shown:

```
Select section object:
Specify a point on the section line to add jog:
```

AutoCAD asks you to specify a point on the section line, and then instantly adds a jog. With each command you can add a single jog a time, but you can add as many jogs as desired. See the following illustration:

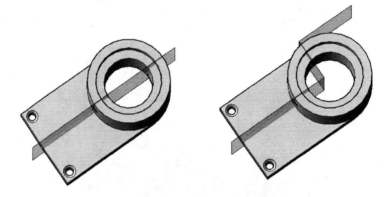

After adding the jog, you can modify it using the grips. See the following illustration:

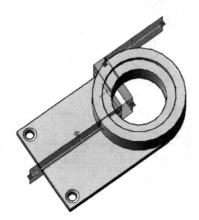

8.7.7 Live Section Command

This command activates the section plane. To issue this command, go to the **Home** tab, locate the **Section** panel, and select the **Live Section** button:

You can also invoke this command by selecting the section plane, right-clicking, and selecting the **Activate live sectioning** option, as shown in the following:

This is the result of activating a section plane:

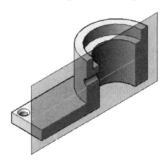

Right-click to select **Show cut-away geometry**. The following is shown:

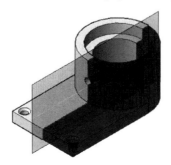

8.7.8 Generate Section Command

This command creates a 2D or 3D section and inserts it as a block in your drawing, or exports it to an external file. To issue this command, go to the **Home** tab, locate the **Section** panel, and select the **Generate Section** button:

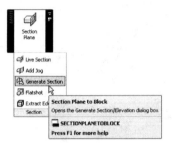

You can also access this command by selecting the section plane, right-clicking, and selecting the **Generate 2D/3D section** option:

The following dialog box is shown:

Click the downward-pointing arrow and the following is displayed:

The top button can be used to select the section plane. Then select whether you want to create a **2D section/Elevation** or **3D section**. Once selected, click the **Create** button. The following prompts are shown:

```
Units: Inches   Conversion:    1.0000
Specify insertion point or [Basepoint/Scale/X/
Y/Z /Rotate]:
Enter X scale factor, specify opposite corner, or
[Corner/XYZ] <1>:
Enter Y scale factor <use X scale factor>:
Specify rotation angle <0>:
```

These are the Insert block command prompts, so there is no need to discuss them further.

Below is an example of a 2D section (the dashed line indicates a hidden line):

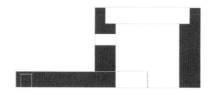

Below is an example of a 3D section:

The following options are available:

- To include all objects or to select part of the drawing to be included in the sectioning process

- To insert the section as a new block, update an existing block (earlier sectioning process), or export the section to a DWG file

- Control the Section Settings; clicking this button displays the following dialog box:

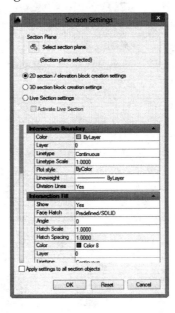

This dialog box controls the settings for 2D sections, 3D sections, and Live sections. You can control the color, linetype, linetype scale, and line-weight of the Intersection Boundary, Intersection Fill, Boundary Lines, and Cut-away Geometry.

PRACTICE 8-3

Using Section Plane Command

1. Start AutoCAD 2015.

2. Open Practice 8-3.dwg.

3. Using the Section Plane command, select the following face:

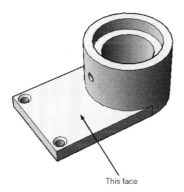

This face

4. Using the Move gizmo, move the section plane upward 13.5 units in the Z direction.

5. Generate a 2D section and insert it as a block in the drawing without changing the settings.

6. Erase the section plane.

7. Create another orthographic section plane, aligned Right.

8. Generate a 2D section, but this time change the Section Settings for Intersection Fill for the Face Hatch to be ANSI31, with Scale = 5, and export it to a file with the name Section Right.dwg.

9. Erase the section plane.

10. Create a section plane using the Draw section option, as in the following:

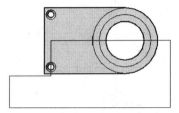

11. Generate a 3D section and export it to a file with the name Section 3D.dwg.

12. Save and close the file.

8.8 SECTIONING USING THE FLATSHOT COMMAND

This command creates a 2D block representing one view of a 3D model.

To issue this command, go to the **Home** tab, locate the **Section** panel, and select the **Flatshot** button:

Before issuing this command, do the following:

- Select the desired view (top, bottom, right, left, etc.)

- To get the right section, make sure you are in *Parallel* mode and not *Perspective* mode. You can do this using the ViewCube.

The following dialog box is displayed:

Most of the dialog box is similar to that described for the Section Plane command. The remaining options are:

- Specify the color and linetype of foreground lines.

- Select whether to show or hide obscured lines, and specify their color and linetype (change them to dashed to simulate correct drafting standards).

When done, click the **Create** button. The following prompts are displayed:

```
Units: Inches   Conversion:   1.0000
Specify insertion point or [Basepoint/Scale/X/
Y/Z/ Rotate]:
Enter X scale factor, specify opposite corner, or
[Corner/XYZ] <1>:
Enter Y scale factor <use X scale factor>:
Specify rotation angle <0>:
```

These are similar to the Insert block command.

Here is an example of the right side of a 3D shape produced by the Flatshot command:

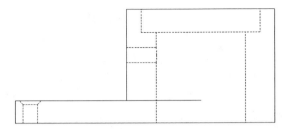

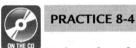

PRACTICE 8-4

Using the Flatshot Command

1. Start AutoCAD 2015.

2. Open Practice 8-4.dwg.

3. Look at the top and right views and generate two flatshots. Don't forget to change the linetype of the obscured lines to Dashed2 and red.

4. Save and close the file.

NOTES

CHAPTER REVIEW

1. Thicken converts a mesh to a solid:

    a. True

    b. False

2. To produce the correct results using the Flatshot command, you should turn the view to _____ and not _____.

3. Which one of the following is not a method to create a section plane?

    a. Selecting a face

    b. Draw section

    c. Parallel to one of the XY, YZ, and ZX planes

    d. Orthographic

4. Convert to Solid converts both meshes and surfaces to solids.

    a. True

    b. False

5. Which one of the following is not a method to create a slice plane?

    a. Planar object

    b. 3 points

    c. Orthographic

    d. Surface

6. Command _____ converts both solids and surfaces to meshes.

CHAPTER REVIEW ANSWERS

    1. b

    3. c

    5. c

CHAPTER 9

PRINTING IN 3D AND 3D DWF

In This Chapter

- Named Views and Viewports in 3D
- Drawing Views
- Generating and Viewing 3D DWF

9.1 NAMED VIEWS AND 3D

You would normally use all 3D viewing commands to look at a model from different angles. When using this command you can save and name views so they can be viewed later or used in layouts for printing purposes. To issue this command, go to the **Visualize** tab, locate the **Views** panel, and select the **View Manager** button:

The following dialog box is displayed:

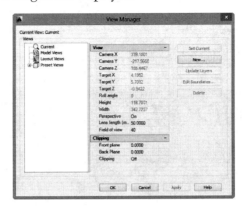

To create a new view, click the **New** button. The following dialog box is displayed:

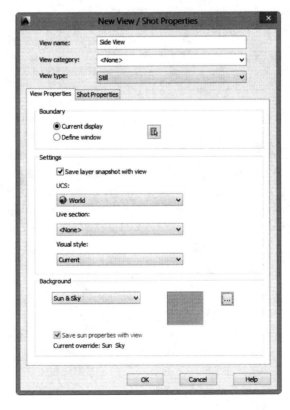

Type in the View's name. Other information that can be saved includes:

- Current layer status
- Current UCS
- Current Live section
- Current visual style

To add a background to the saved view, pick one of the five available choices:

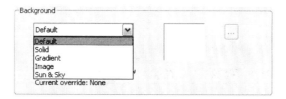

A dialog box is displayed that allows you to manipulate the parameters of the five selections. When you are done, click OK to execute the command.

The following picture is displayed. This dialog box presents a list of saved views:

Another list of saved views is available in the Views panel. See the following picture (save the file first):

9.2 3D AND VIEWPORT CREATION

This section discusses the special features of 3D in viewport creation. To issue this command, go to the **Visualize** tab, locate **Model Viewports**, click the **Named** button, and click the **New Viewports** tab. The following dialog box is displayed:

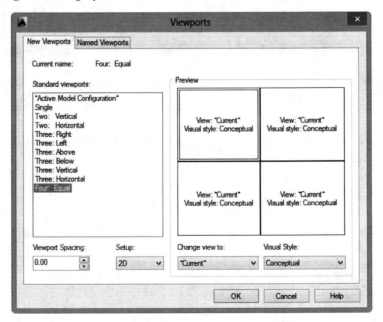

Changing the **Setup** allows you to choose 2D or 3D. If you select 3D, the arrangement under **Preview** changes to the following:

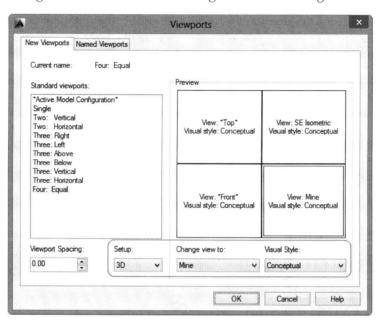

In the image above you can see that AutoCAD assigns a view for each viewport. You can change the default view and assign a saved view for viewports in the **Change view to** list. You can change the Visual style as well. The result is something similar to the following:

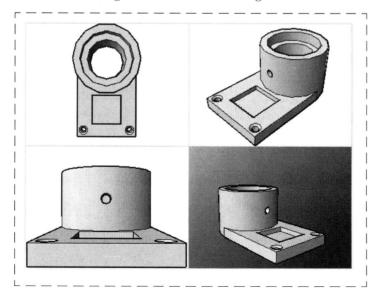

A shade mode can be assigned to a viewport. Select the desired viewport, right-click, select **Shade plot**, and select one of the modes, as shown below:

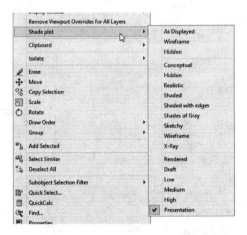

Note that the bottom of the shade plots list includes rendering options. You can plot a rendered image if you want, but the effects of this action are only revealed by using the **Print Preview** command.

PRACTICE 9-1

ON THE CD

View and Viewport with 3D

1. Start AutoCAD 2015.

2. Open Practice 9-1.dwg.

3. Change the visual style to Hidden.

4. Using the different viewing commands, try to achieve the following view, or something close to it:

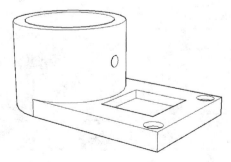

5. Save this view under the name "My First View."

6. Change the background to Gradient and set the angle to 45º.

7. Go to Layout1.

8. From the **Visualize** tab, locate the **Model Viewports** panel. Then, using the **Named** button, click the **New Viewports** tab.

9. Select Four: Equal, change the setup to 3D, select the lower right viewport, and change the view to My First View. Make sure the visual style for the other three viewports is Conceptual.

10. Insert the four viewports in the space available in Layout1.

11. Select the top right viewport and change the shade plot to Presentation.

12. Using Plot Preview, check the output. You should have something similar to the following:

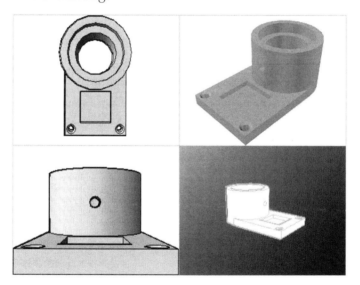

13. Save and close the file.

9.3 MODEL DOCUMENTATION

Model Documentation generates intelligent documentation for AutoCAD 3D models. Model Documentation generates views, sections, and details that are instantly updated when the model changes. To work with this tool, go to a layout and all the panels are in the **Layout** tab (displayed only in Layout, NOT in the model space). These are Create View, Modify View, Update, and Styles and Standards:

The procedure is very simple:

- Create a Base view

- Create Projected views

- Create Sections and Details

- Edit and update views

- Control Styles and Standards

9.4 CREATING A BASE VIEW

This should always be your first step. If you have more than one 3D model, select the objects to be included during the creation process. You can start this process from the Model space or your desired layout. To start this command, go to the **Layout** tab, locate the **Create View** panel, and select the **Base** button:

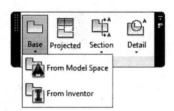

There are two options from which to create a Base view:

- Current Model Space

- Inventor; this option allows the import of an Inventor file (*.iam, *.ipt, or *.ipn)

To include 3D objects created in the Model Space, there are two options:

- Start the command in Model Space. AutoCAD asks you to select objects to include in the view/section/detail, or you can select the entire model. Because you are in the Model Space, AutoCAD asks you specify the name of the layout to be used. The following prompts are displayed:

```
Select objects or [Entire model] <Entire
model>: Enter new or existing layout name to
make current or [?] <Layout1>: layout1
```

- Start the command in one of the layouts. AutoCAD asks you to specify the location of the Base view. The Select option is the next important step. By default, AutoCAD selects all objects in the Model Space. When you start the Select option you are taken to the Model space to remove the undesired objects. When done, AutoCAD takes you back to layout you were in. The following prompts are displayed:

```
Type = Base only  Hidden Lines = Visible and
hidden lines  Scale = 1/8" = 1'-0"Specify
location of Base view or [Type/sElect/
Orientation/Hidden lines/Scale/Visibility]
<Type>:
```

A new context tab named **Drawing View Creation** is displayed, as follows:

The Model Space Selection button on the left works just like the Select option discussed above.

The Default view appears. You can change your Base view from the default view using two methods. The first is to use the **Orientation** panel in the context tab. The second is to use the **Orientation** option in the prompts mentioned above.

By default, using the Base command inserts the Base view and Projected views. Use the Type option to insert the Base view or the Base and Projected view.

Use the **Appearance** panel in the context tab to determine the Hidden Lines settings. There are four options: Visible lines, Visible and hidden lines, Shaded with visible lines, and Shaded with visible and hidden lines. Use the same panel to set the scale and edge visibility.

Assuming you have inserted the Base view only, you will see something similar to the following:

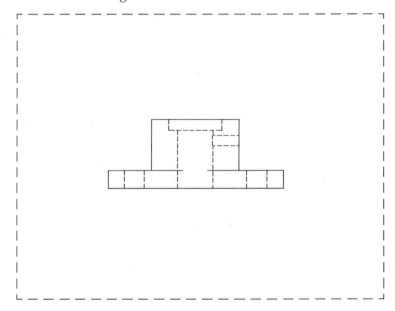

9.5 CREATING PROJECTED VIEWS

If a Base view is created, but not a projected view (as in the previous step), they can be created afterward. To insert a projected view, go to the **Layout** tab, locate the **Create View** panel, and select the **Projected** button:

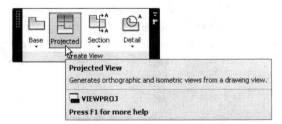

The following prompts are displayed:

```
Select parent view:
Specify location of projected view or <eXit>:
Specify location of projected view or [Undo/eXit]
<eXit>:
```

Select the parent (base) view and then create projected views from it. You can create up to eight views; four are orthographic, and four are isometric. You will see something similar to the following:

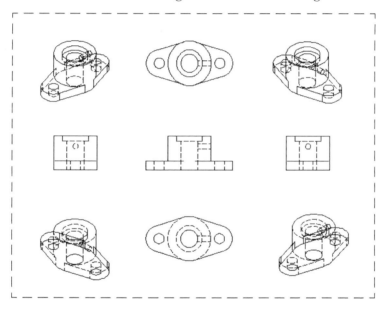

9.6 CREATING SECTION VIEWS

After inserting the Base view, AutoCAD can produce section views. To use the section commands, go to the **Layout** tab, locate the **Create View** panel, and click the **Section** list. Something similar to the following is displayed:

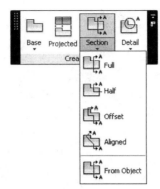

There are five ways to create a section from a base view. These are:

- Full Section
- Half Section
- Offset Section
- Aligned Section
- Section from Object

Using the Full section displays the following prompts:

```
Specify start point:
Specify end point or [Undo]:
Specify location of section view or:
```

Specify the start and end points of the section line, and specify the location of the section. You will see something similar to the following:

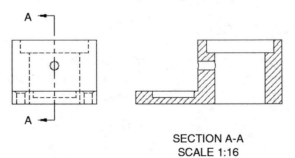

SECTION A-A
SCALE 1:16

A half section is a view of an object that displays half of the view. The following prompts are displayed:

```
Specify start point:
Specify next point or [Undo]:
Specify end point or [Undo]:
Specify location of section view or:
```

Specify three points, and specify the location of the section. You will see something similar to the following:

SECTION A-A
SCALE 1:16

An offset section is a section view along a line that does not follow a straight line. The following prompts are displayed:

```
Specify start point:
Specify next point or [Undo]:
Specify next point or [Undo]:
Specify next point or [Undo]:
Specify next point or [Undo/Done] <Done>:
```

You will see something similar to the following:

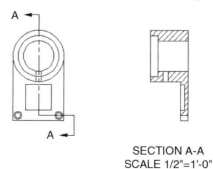

SECTION A-A
SCALE 1/2"=1'-0"

An aligned section creates a section line on an angle. The following prompts are displayed:

```
Specify start point:
Specify next point or [Undo]:
Specify next point or [Undo]:
Specify next point or [Undo/Done] <Done>:
```

You will see something similar to the following:

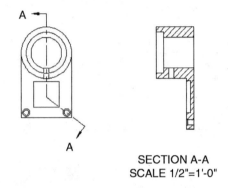

SECTION A-A
SCALE 1/2"=1'-0"

Lastly, an existing object can serve as a section line. The following prompts are displayed:

```
Select objects or [Done] <Done>:
Select objects or [Done] <Done>:
Specify location of section view or:
```

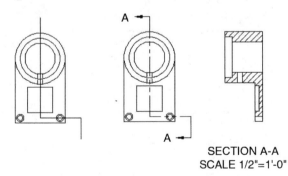

SECTION A-A
SCALE 1/2"=1'-0"

Using any of the following methods displays a context tab titled **Section View Creation**. It looks similar to the following:

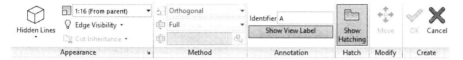

You can control the following:

- Appearance (Hidden Lines, Scale, Edge Visibility)
- Annotation (Identifier, show or hide label)
- Hatch (show or hide hatch)

9.7 SECTION VIEW STYLE

The Section View Style controls the view styles in the section. To activate this command, go to the **Layout** tab, locate the **Styles and Standards** panel, and click the **Section View Style** button:

The following dialog box is displayed:

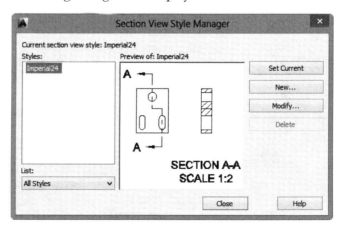

You can create a new style or modify an existing one (Imperial24 is an existing style). If you click the **New** button, the following dialog box is displayed:

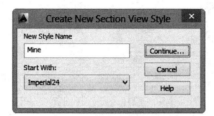

Input the name of the new style and click the **Continue** button. The following is displayed:

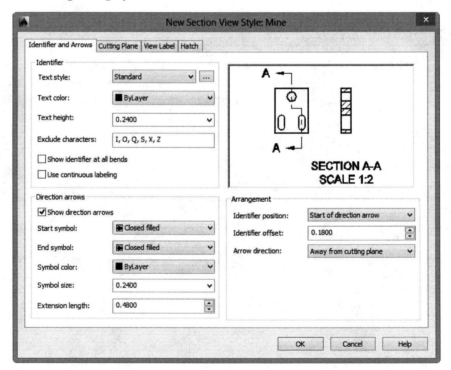

There are four tabs in this style dialog box. They are:

- Identifier and Arrows
- Cutting Plane
- View Label
- Hatch

Here is an explanation of each:

9.7.1 Identifier and Arrows Tab

You will see the following:

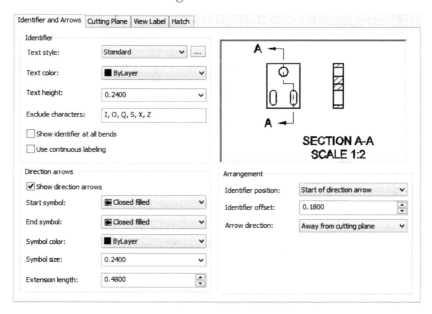

In this tab, you can:

- Specify Text style, Text color, and Text height
- Specify characters to be excluded from identifier names (the default excludes I, O, Q, S, X, Z)
- Show or hide identifiers at all bends of the section line
- Show or hide continuous labeling

- Show or hide arrow directions

- Specify the start and end symbols

- Specify Symbol color and size

- Specify Extension length of the line before the arrow

- Specify the identifier position (five options):

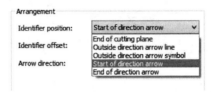

- Specify the Identifier offset (the distance before inserting the identifier)

- Specify the Arrow direction (two choices):

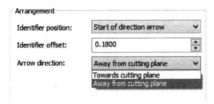

9.7.2 Cutting Plane Tab

You will see the following:

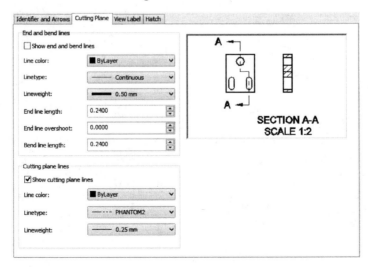

In this tab, you can:

- Show or hide end and bend lines
- Specify Line color, Linetype, Lineweight
- Specify end line length, which sets the length of the end line segment
- Specify end line overshoot, which sets the length of the extension beyond the direction arrows
- Specify the bend line length, which sets the length of the bend line segment on either side of the bend vertices
- Show or hide cutting plane lines, and set their color, linetype, and lineweight

9.7.3 View Label Tab

You will see the following:

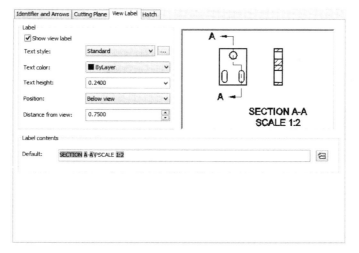

In this tab, you can:

- Show or hide the view label
- Specify Text style, Text color, and Text height
- Specify the position of the label (above or below)
- Specify the distance of the label from the view
- Specify the default content

9.7.4 Hatch Tab

You will see the following:

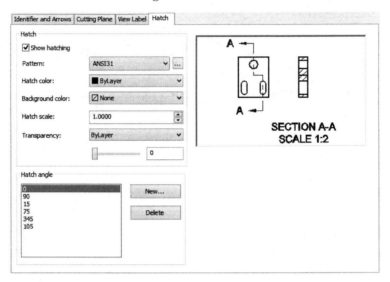

In this tab, you can:

- Show or hide hatch

- Specify hatch pattern, hatch color, hatch background color, hatch scale, and hatch transparency (either leave it as ByLayer or specify a value)

- Specify hatch angle (use one of the existing values or define your own using the New button)

9.8 CREATING DETAIL VIEWS

AutoCAD can produce detail views to magnify small details. To use the detail commands, go to the **Layout** tab, locate the **Create View** panel, and click the **Detail** list. You will see something similar to the following:

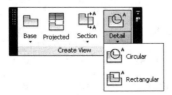

There are two options:

- Circular
- Rectangular

The following prompt is displayed for both:

```
Specify center point or [Hidden lines/Scale/
Visibility/Boundary/model Edge/Annotation]
<Boundary>
```

Here is an example of a circular detail view:

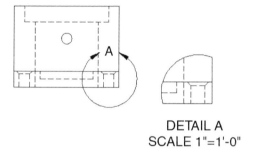

DETAIL A
SCALE 1"=1'-0"

Here is an example of a rectangular detail view:

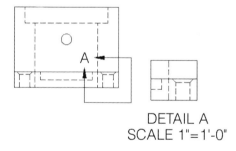

DETAIL A
SCALE 1"=1'-0"

Using any of the following methods displays a context tab titled **Detail View Creation**. It looks like the following:

You can control the following:

- Appearance (Hidden Lines, Scale, Edge Visibility)
- Boundary (the type selected is the type you started with)
- Model edge (four options: Smooth, Smooth with border, Smooth with connection line, and Jagged)
- Annotation (Identifier, show or hide label)

9.9 DETAIL VIEW STYLE

Detail View Style controls the appearance of everything created in the detail view. To activate this command, go to the **Layout** tab, locate the **Styles and Standards** panel, and click the **Detail View Style** button:

The following dialog box is displayed:

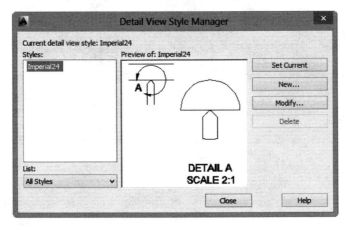

You can create a new style or modify an existing one (Imperial24 is an existing style). If you click the **New** button, the following dialog box is displayed:

Input the name of the new style and click the **Continue** button to edit the style. Three tabs are displayed:

- Identifier
- Detail Boundary
- View Label

9.9.1 Identifier Tab

You will see something similar to the following:

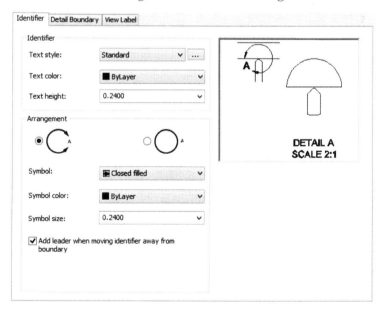

In this tab, you can:

- Specify the text style, text color, and text height for the identifier

- Select one of the two arrangements available

- Select the symbol, symbol color, and symbol size, and whether or not a leader is added when moving the identifier away from the boundary

9.9.2 Detail Boundary Tab

You will see something similar to the following:

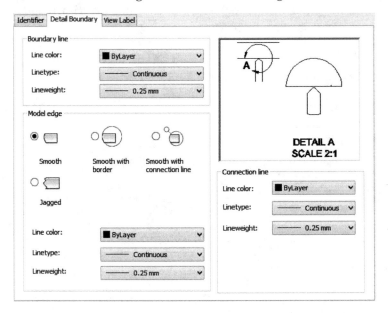

In this tab, you can:

- Specify a boundary line's color, linetype, and lineweight

- Specify the model edge (select one of the four available methods), and that line's color, linetype, and lineweight

- Specify if there is a connection line, and that line's color, linetype, and lineweight

9.9.3 View Label Tab

You will see something similar to the following:

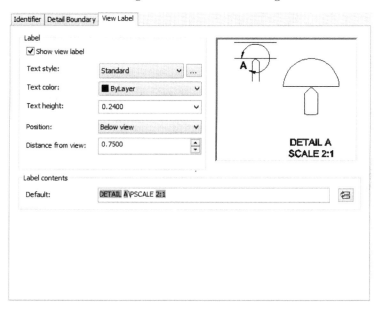

In this tab, you can:

- Show or hide the view label
- Specify the Text style
- Specify the Text color and Text height
- Specify the position of the label (above or below)
- Specify the distance of the label from the view
- Specify the default content

9.10 EDITING VIEW

AutoCAD allows you to edit a view after it is inserted. A special context tab is displayed depending on the type of view. To activate the Edit View command, go to the **Layout** tab, locate the **Modify View** panel, and select the **Edit View** button:

The following prompt is displayed:

```
Select view:
```

Select the desired view (base, section, or detail). For a base view, the Drawing View Editor context tab is displayed:

These buttons have been discussed in the section on the creation process. For a section view, the Section View Editor context tab is displayed:

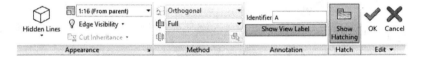

These buttons have been discussed previously. For a detail view, the Detail View Editor context tab is displayed:

These buttons have been discussed previously.

Some things to consider:

- The same effect takes place if you double click the desired view.

- A Projected view displays the same Drawing View Editor context tab, but the selection button is off.

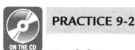

PRACTICE 9-2

Model Documentation

1. Start AutoCAD 2015.

2. Open Practice 9-2.dwg.

3. You will see two parts.

4. Go to Part (1) layout.

5. Go to the Layout tab and locate the Create View panel. Select the Base button, and select From Model Space.

6. You can see the two parts that are selected. Right-click and choose the Select option, which takes you to the Model Space. Select the remove option, and select the part on the left to be removed. Press [Enter].

7. Using the context tab, make sure that Front and Hidden Lines are selected, and set the scale to ½" = 1'.

8. Insert the view at the middle center of the sheet, and one projected view to the left. Then press [Enter] to execute the command.

9. Start Section View Style and modify the existing style using the following:

   a. Text height for identifier = 0.14

   b. Under the Direction arrows, the start and end symbols should be blank.

   c. Symbol size = 0.12

   d. Text height for view label = 0.14, and distance from view = 0.4

   e. Under Hatch, Hatch scale = 0.75

10. Using Full Section, add a section to the right of the base view, cutting the middle of the base view (use OTRACK and the circle in the middle).

11. Start Detail View Style and modify the existing style using the following:

    a. Text height for identifier = 0.14

    b. Arrangement = Full circle

    c. Under Detail Boundary, select Smooth with connection line.

    d. Text height for view label = 0.14, and distance from view = 0.4

12. Add a circular detail for the screw hole on the right, below the base view using a scale of 1" = 1' (if the text extends beyond the dashed line, you can move it up).

13. Turn off the Auto Update button.

14. Go to Model Space.

15. Click the Home button of the ViewCube.

16. Zoom to the top of the shape on the right, and fillet the outside edge of the top using radius = 1.5.

17. Go to Part (1) layout.

18. Select to update all views.

19. Zoom to the section on the right.

20. Make MD_Annotation the current layer.

21. Using the normal dimension, put the following dimensions:

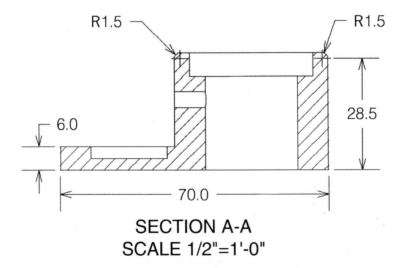

SECTION A-A
SCALE 1/2"=1'-0"

22. Go back to Model Space and delete the fillet you added in step 16.

23. Go to Part (1) layout and update all views.

24. Note the alert badges on the drawing and the bubble that is displayed on the Annotation Monitor. Close the bubble without using the link.

25. Manually delete the two radius dimensions.

26. Reassociate the linear dimension on the right with the full length.

27. Go to Part (2) layout and create a similar layout to that on the left using a scale of ¼" = 1', and select the proper scales for the other views.

28. Save and close the file.

9.15 3D DWF CREATION AND VIEWING

3D DWF files can be viewed in Autodesk Design Review software with high intelligence. 3D DWF allows you to look at the 3D model from different angles, dismantle the 3D model, and create sections of the model. There are two ways to produce a 3D DWF:

- Single-sheet 3D DWF using the 3D DWF command
- Multiple-sheet 3D DWF using the Batch Plot command

9.15.1 3D DWF Command

This command produces a single-sheet 3D DWF file. To issue this command, go to the **Output** tab, locate the **Export to DWF/PDF** panel, and select the **3D DWF** button:

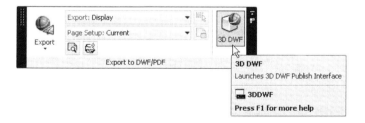

The following dialog box is displayed:

Once AutoCAD has finished producing the file, the following message is displayed:

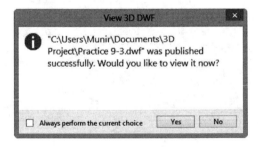

If you select Yes, AutoCAD opens Autodesk Design Review to view the 3D file.

9.15.2 Batch Plot Command

This command produces a multiple-sheet 3D DWF file. To issue this command, go to the **Output** tab, locate the **Plot** panel, and select the **Batch Plot** button:

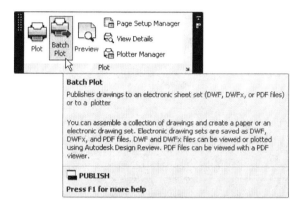

The following dialog box is displayed:

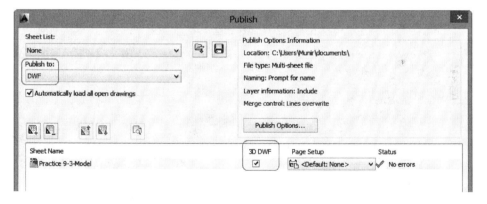

Under **Publish to**, select DWF or DWFx. For a Model sheet, turn on the checkbox for 3D DWF. This is the only way to inform AutoCAD that you want 3D DWF. To control the 3D DWF, click the **Publish Options** button and the following dialog box is displayed:

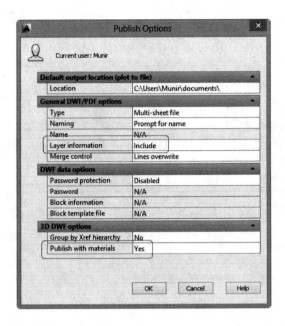

Set the **Layer information** value to **Include**. At the bottom of the dialog box, set the **Publish with materials** value to **Yes**.

Note that not all data of DWG is transferred to 3D DWF. Data not transferred includes:

- Animation

- Some Fonts

- Lights and Shadow

To see a full list, see AutoCAD Help and the manual.

NOTE *Autodesk Design Review 2013 is in use, as Autodesk has not updated the software with the newer versions of AutoCAD. You can download the software from Autodesk.com free-of-charge.*

After completing production of the 3D DWF, start Autodesk Design Review for viewing. Something similar to the following is displayed:

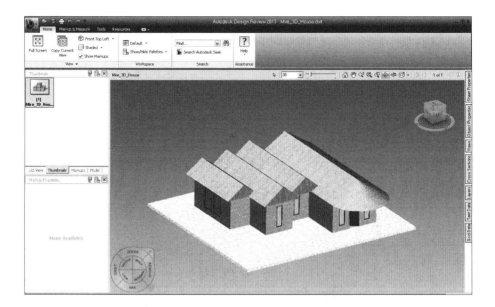

On the left side of the screen is a small window named **Model**. This window contains all the layers in your original AutoCAD file. Clicking one of the layers lists the solids in it. This is a good method to select solids if you don't want to use the mouse to pick up objects.

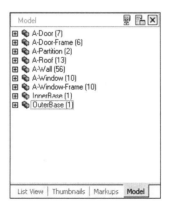

Note the ViewCube in the top right corner of the screen. Other viewing commands are shown in the top canvas. If you select the **Tools** tab, two active panels are displayed: **3D Tools** and **Create Sheet.** The following options are offered:

- Move & Rotate

- Sectioning buttons

- Create a sheet in the DWF file from snapshot

9.15.3 Move & Rotate Command

This command takes the model apart by moving and rotating different parts of the model. To issue this command, go to the **Tools** tab, locate the **3D Tools** panel, and select the **Move & Rotate** button:

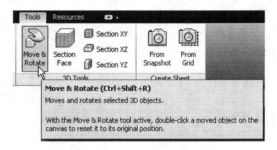

When this command is active the cursor becomes the shape of a human hand. When the hand hovers over any 3D object, it is highlighted, and you can select the object with a simple click. The desired object is similar to the UCS icon. Using this icon you can move in X, Y, Z, XY, YZ, and ZX directions, and rotate around X, Y, and Z. See the following illustrations:

- Moving along Z-axis:

CAMERAS AND LIGHTS

In This Chapter

- Creating and Controlling Cameras
- Creating and Editing Various Types of Lights

10.1 INTRODUCTION

Chapter 10 and Chapter 11 are closely connected, as mastering the concepts presented in these two chapters will lead to the production of professionally rendered images. This chapter discusses the first two steps:

- Creating, setting up, and editing cameras in the drawing
- Creating, setting up, and editing lights in the drawing

Once these two topics are mastered you will be ready to tackle Chapter 11, which deals with materials and the production of rendered images.

10.2 CREATING A CAMERA

Creating a camera in AutoCAD involves some research to achieve the best results. Specify two positions in the XYZ space: one for the camera and one for the target. First establish the limits of your model in the XY plane, and then determine the best Z value from which to view your model. Osnap can also be used to specify the two positions. When the creation process is

successful, the Field-of-View (FOV) is created based on the camera lens
length. See the illustration below:

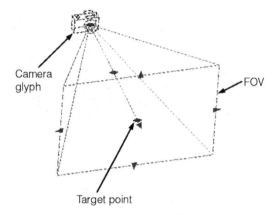

To create a new camera, go to the **Visualize** tab, locate the **Camera**
panel, and select the **Create Camera** button:

The following prompts are displayed:

```
Current camera settings: Height=0 Lens
Length=50.0 mm
Specify camera location:
Specify target location:
Enter an option [?/Name/LOcation/Height/Target/
LEns/ Clipping/View/eXit]<eXit>:
```

The first line presents the current value for camera height and lens
length. AutoCAD then asks you to specify the coordinates of both the
camera and the target points. You can type the coordinates or specify points
on the screen using Osnap. When you are done, you can exit the command
or change one of the following parameters:

- Name: changes the name of the camera to something other than
  the default name of Camera 1.

- Location: changes the location of the camera point.

- Height: changes the camera height.

- Target: changes the target position.

- Lens: changes the lens length of the camera. The greater the lens length, the smaller the FOV.

- Clipping: discussed later in this chapter.

- View: switches to the camera view.

- Exit: exits the command.

10.3 CONTROLLING A CAMERA

Once the camera is created, a camera glyph is displayed in the drawing. Clicking the camera glyph does two things:

- A **Camera Preview** window box appears that displays the camera view. You can select the visual style to be used.

- Grips appear, showing the camera, target points, and the FOV.

10.3.1 Camera Preview Dialog Box

Click the camera glyph in the drawing and the **Camera Preview** dialog box appears and displays the current camera view. In the dialog box, you can change the visual style and specify whether or not this dialog box is displayed in the future. See the illustration:

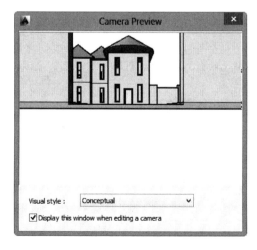

10.3.2 Grips

Selecting the camera glyph displays grips and a pyramid with a rectangular base and an apex point. The apex point is the camera point, and the center of the rectangular base is the target point. The pyramid base represents the FOV. The following picture illustrates this concept:

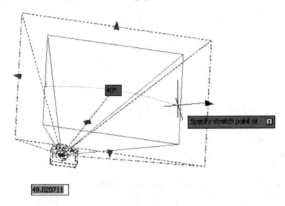

Select the camera, right-click, and select the **Properties** option on the **Properties** palette of the camera. The following is displayed:

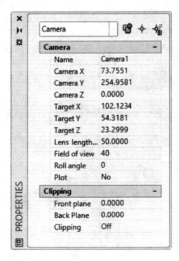

All camera properties can be changed here, from the camera name to enabling the front and back clipping planes.

10.3.3 Front and Back Clipping Planes

Front and back clipping planes hide some or all of the view depending on where they are positioned. Objects between the camera position and the front clipping plane are not visible, and objects between the target and the back clipping plane are not visible.

You can turn the clipping panes on and off in the **Properties** palette, as shown in the following picture:

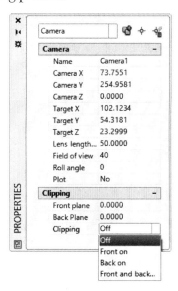

Moving the front clipping plane away from the camera obscures some or all of the model, and moving the back clipping plane closer to the camera also obscures some or all of the model. It is advised that you change to the top view for clipping. Using the top view allows you to identify both the front clipping plane and back clipping plane easily. The following illustration explains this concept:

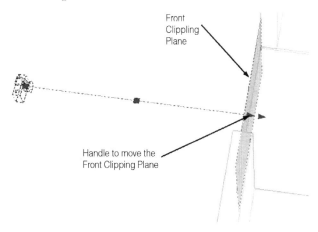

By moving one of the two clipping planes and using the Camera Preview dialog window, you can see how clipping affects the scene.

If a clipping plane is enabled its effects are displayed. To remove clipping effects, turn off the clipping plane.

10.3.4 Camera Display and the Camera Tool Palette

The camera glyph is shown by default. To hide it, go to the **Visualize** tab, locate the **Camera** panel, and select the **Show Camera** button. This button turns the camera glyph on and off:

AutoCAD has predefined cameras that are ready for use. They can be accessed through the Tool Palettes. The palette called **Cameras** features three cameras. They are:

- Normal Camera (50 mm lens length)

- Wide-angle Camera (35 mm lens length)

- Extreme Wide-angle Camera (6 mm lens length)

See the following illustration:

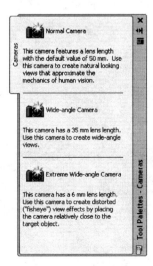

To issue this command, go to the **Visualize** tab, locate the **Lights** panel, and select the **Spot** button:

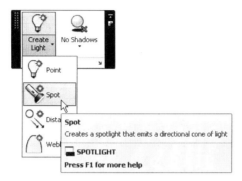

Two cones work together to manage how the light diminishes. These are:

- Hotspot Cone
- Falloff Cone

Light intensity is consistent in the Hotspot cone, but diminishes from the perimeter of the Hotspot cone to the perimeter of the Falloff cone.

The following picture illustrates this concept:

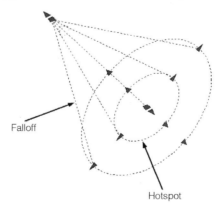

The following prompts are displayed:

```
Specify source location <0,0,0>:
Specify target location <0,0,-10>:
Enter an option to change [Name/Intensity factor
/Status/Photometry/ Hotspot/Falloff/shadoW/
Attenuation /filterColor/eXit]
```

Specify a source position in XYZ, and a target position. The remaining options are discussed in the section on the Point light command, except for the following two options:

10.6.1 Hotspot Option

Specify the angle of the Hotspot cone. The following prompt is displayed:

```
Enter hotspot angle (0.00-160.00) <45.0000>:
```

As the prompt indicates, input a value between 0° and 160°.

10.6.2 Falloff Option

Specify the angle of the Falloff cone. The following prompt is displayed:

```
Enter falloff angle (0.00-160.00) <50.0000>:
```

As the prompt indicates, input a value between 0° and 160°.

To access the Free Spot light, type **light** in the command window. The following prompts are displayed:

```
Specify source location <0,0,0>:
Enter an option to change [Name/Intensity factor/
Status/ Photometry/Hotspot/Falloff/shadoW/
Attenuation/filterColor/eXit]
```

10.7 INSERTING DISTANT LIGHT

Distant light, like the other sources of light, has a source location and a target location. Distant light emits light along a line parallel to the line between the source and target points. Only objects in the path of light are affected. To issue this command, go to the **Visualize** tab, locate the **Light** panel, and select the **Distant** button:

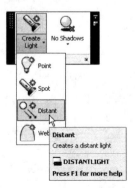

The following prompts are displayed:

```
Specify light direction FROM <0,0,0> or [Vector]:
Specify light direction TO <1,1,1>:
Enter an option to change [Name/Intensity factor /
Status/Photometry/shadoW/filterColor/eXit] <eXit>:
```

Specify the positions of the source and the target. The remaining options have already been discussed.

10.8 DEALING WITH SUNLIGHT

You can see the effect of sunlight on your 3D model so you can produce realistic rendering. To issue your geographic location, go to the **Visualize** tab, locate the **Sun & Location** panel, and select the **Set Location** button. You can set location from a map or from a file:

Set your location by specifying a GIS file such as *.kmz, or *.kml, or use the map provided by Autodesk. To use the map you should have an account with Autodesk 360 (Autodesk cloud); sign in before setting the location.

The following dialog box is displayed:

Type the name of the desired city into the search field and click the search button. You will see a view of the entire city. By zooming and panning, locate the desired location in your city, and click the ***Drop Marker Here*** button on the left. A small, red push pin is displayed. Click Continue to specify the coordinates of the location and the direction north.

Next, change the sun status to ON by going to the **Visualize** tab, locating the **Sun & Location** panel, and selecting the **Sun Status** button:

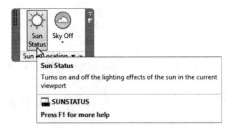

To see the effect of the sun at different times of the year and different hours of the day, control the **Date** and **Time** in the same panel by dragging the two sliders, as shown:

You can add a **Sky effect** if it is an outside scene. There are three options:

AutoCAD allows you to control the **Sun Properties** by using the same panel and clicking the small arrow on the right, as shown:

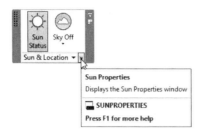

The sun palette is displayed for editing:

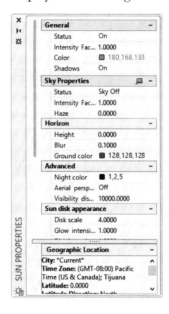

10.9 INSERTING WEBLIGHT

All previously discussed light types are designed to produce light in a consistent fashion, but not weblight. Weblight is designed to produce non-uniform light. Using weblight helps AutoCAD produce rendered images that look and feel like the real world. To help clarify this concept, imagine a sphere with point light distributed in a certain way (not necessarily covering the entire surface of the sphere).

This is the weblight, and the distribution of light is determined by a file provided by the light's manufacturer. AutoCAD provides two types of weblight:

- Weblight: has a source and target points.

- Free Weblight: does not have a target point.

To issue this command, go to the **Visualize** tab, locate the **Light** panel, and select the **Weblight** button:

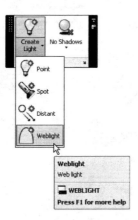

The following prompts are displayed:

```
Specify source location <0,0,0>:
Specify target location <0,0,-10>:
Enter an option to change [Name/Intensity factor/
Status/Photometry/weB/shadoW/filterColor/eXit]
<eXit>:
```

All prompts have been covered except for the following option:

10.9.1 Web Option

Choosing this option displays the following prompt:

```
Enter a Web option to change [File/X/Y/Z/Exit]
<Exit>:
```

AutoCAD asks you to type in the coordinates of the lights around the sphere (most likely this is a very lengthy and tedious job), or load a file from the manufacturer that has the coordinates. Use the **Properties** palette to load your desired file.

NOTES

CHAPTER REVIEW

1. FOV means Field of View.

    a. True

    b. False

2. Which one of the following statements is incorrect?

    a. There are two types of Spot light.

    b. You cannot see the effects of light in Conceptual visual style.

    c. There is one type of Distant light.

    d. You can see the effects of light in Conceptual visual style.

3. Which one of the following statements is not true?

    a. There are two cones for Spot light.

    b. The greater the lens length, the smaller the FOV.

    c. FOV is a pyramid, with the camera point as the apex.

    d. To set up your location, it is not a condition to have an account with Autodesk 360.

4. Spot light and No-Target spot lights are the two types of spot lights that AutoCAD supports.

    a. True

    b. False

5. The file extension for a Weblight file is _____.

6. Extreme Wide-angle camera uses _____ mm lens length.

CHAPTER REVIEW ANSWERS

1. a

3. d

5. IES

MATERIAL, RENDERING, VISUAL STYLE, AND ANIMATION

In This Chapter

- Loading and Assigning Materials to Solids and Faces
- Material Mapping
- Creating Your Own Material
- Rendering and Rendering Region
- Creating Your Own Visual Style
- Creating Animation

11.1 INTRODUCING MATERIALS IN AutoCAD

Using materials in AutoCAD involves several steps. These are:

- Selecting the desired material library from Material Browser
- Dragging the desired materials to be used
- Assigning materials to objects or faces
- Assigning material mapping to different objects and faces
- Rendering to see the effects of your assignments

AutoCAD has a premade library of materials that includes Ceramic, Concrete, Flooring, and more. To achieve the best results you can assign material mapping for each face or object. Material mapping tells AutoCAD how to repeat the pattern, and which angle to use.

11.2 MATERIAL BROWSER

To start working with materials in AutoCAD, start the Material Browser, which includes all material definitions. To issue this command, go to the **Visualize** tab, locate the **Materials** panel, and select the **Material Browser** button:

The Material Browser palette is displayed as shown below:

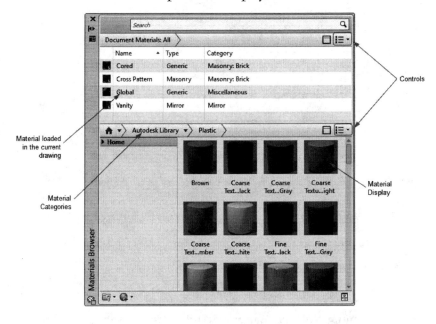

As shown, the Material Browser has three parts:

- The top part shows material loaded in the current drawing
- The left part shows the existing material libraries (one of them is the Autodesk Library, as shown), and the categories holding the materials
- The right parts shows the material display

First, load materials from the library to your drawing by dragging the desired material from the library and dropping it onto the top portion of the material browser.

The two controls on the right control how the different parts are displayed. The following depicts the control for the top part of the Material Browser (it is self-explanatory):

This is the control for the materials at the bottom of the Material Browser:

11.3 ASSIGNING MATERIALS TO OBJECTS AND FACES

There are three methods to assign materials to objects and faces. They are:

- Drag-and-drop

- After selecting an object, right-click one material

- Using the Attach by layer command

11.3.1 Drag-and-Drop

While the Material Browser is displayed, select the library, category, and material from the right section. Select the material and drag-and-drop it on the desired object (solid, surface, or mesh). To assign the material to a single face, hold the [Ctrl] key when assigning; the material is assigned only to the selected face.

11.3.2 After Selecting an Object, Right-Click Material

This method also includes the selection of objects (or faces), as in the first step. After the material is loaded into the current file (using the drag-and-drop method), right-click it. The following menu is displayed. Select the **Assign to Selection** option, and the material is assigned:

11.3.3 Attach by Layer Command

For this method to work properly, make sure all 3D objects are in the correct layer. If this is the case, issue the command by going to the **Visualize** tab, locating the **Materials** panel, and selecting the **Attach By Layer** button:

The following dialog box is shown. It includes the material loaded in the current drawing on the left, and the layers on the right:

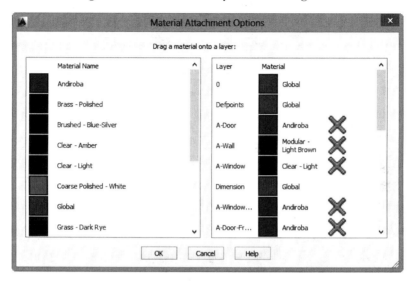

Drag the material from the left side and drop it on the layer on the right side.

11.3.4 Removing Assigned Materials

This command removes the material assigned to an object. To issue this command, go to the **Visualize** tab, locate the **Materials** panel, and select the **Remove Materials** button:

The following prompt is displayed:

```
Select objects:
```

The mouse changes to the following shape:

Select the desired objects. When done, press [Enter] to execute the command.

11.4 MATERIAL MAPPING

Material mapping is defined by the following:

- The shape of your object (planar, cylindrical, or spherical)
- The number of times your image is repeated in columns and rows
- The value of the angle of the image
- The location of the image to start repeating

With material mapping, AutoCAD produces more realistic rendered images. To issue this command, go to the **Visualize** tab, locate the **Materials** panel, and select the **Material Mapping** button:

As shown, there are four types of material mapping:

- Planar
- Box
- Cylindrical
- Spherical

11.4.1 Planar Mapping

This option works for planar faces and objects, but it is specifically intended for faces. The following prompt is displayed:

```
Select faces or objects:
```

Select the desired face. A green frame with a triangle at each corner is displayed:

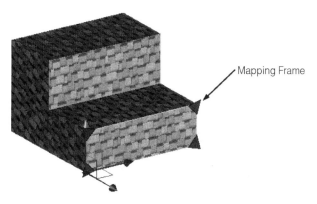

Mapping Frame

The following prompt is then displayed:

```
Accept the mapping or [Move/Rotate/reseT/sWitch
mapping mode]:
```

You can resize, move, and rotate the mapping. The following example illustrates resizing the mapping:

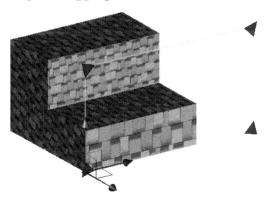

Compare the resized pattern to the other parts of the object.

See the example below, illustrating rotation:

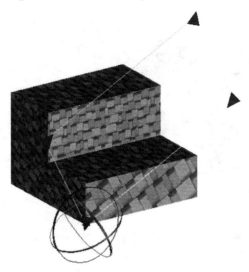

Note the pattern and compare it to the other parts of the object. When done, press [Enter] to end the mapping process.

11.4.2 Box Mapping

Box mapping is designed for box-like objects. The following prompts are displayed:

```
Select faces or objects:
```

A mapping box similar to the following is displayed:

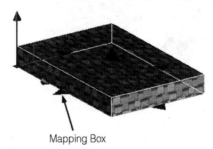

Mapping Box

Five triangles control the size. Use these triangles to resize, rotate, and move the box mapping. The following prompt is displayed:

```
Accept the mapping or [Move/Rotate/reseT/sWitch
mapping mode]:
```

This same prompt has been discussed in the section on planar mapping. See the following example:

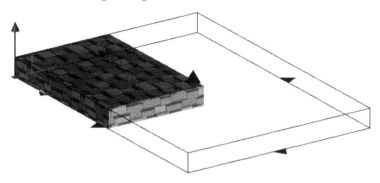

Compare this picture to the picture prior to box mapping. When done, press [Enter] to end the command.

11.4.3 Cylindrical and Spherical Mapping

These two mapping methods are designed for cylindrical and spherical shapes. They use the same prompts. The following pictures illustrate this concept:

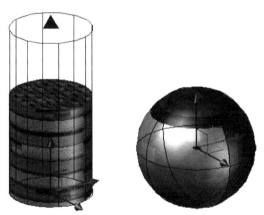

Note that cylindrical mapping has a single triangle with which you can increase or decrease the mapping along the height, but spherical mapping does not. Spherical mapping only provides the ability to move the pattern toward the three main directions.

11.4.4 Copying and Resetting Mapping

To expedite the mapping process, you can do two things:

- Copy existing mapping from one object to another.

- Reset mapping to the default to start your mapping again.

To issue these two commands, go to the **Visualize** tab, locate the **Materials** panel, and select one of the two buttons shown:

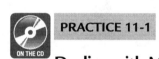

PRACTICE 11-1

Dealing with Materials

1. Start AutoCAD 2015.

2. Open practice 11-1.dwg.

3. Load the following materials from the Material Browser:

Category	Material
Wood	Andiroba
Metal	Brass-Polished Brushed-Blue-Silver
Glass	Clear-Amber Clear-Light
Stone	Coarse Polished-White
Sitework	Grass-Dark Rye
Flooring	Maple-Rosewood
Masonry	Modular-Light Brown
Fabric	Plaid 4 Stripes-Yellow-White

4. Assign materials to layers according to the following table:

Material	Layer
Andiroba	A-Door/A-Door-Frame/A-Window-Frame
Brass-Polished	Table_legs
Brushed-Blue-Silver	Lamp_Rod
Clear-Amber	Table_Top
Clear-Light	A-Window
Coarse Polished-White	Lamp_Base
Grass-Dark Rye	OuterBase
Maple-Rosewood	InnerBase
Modular-Light Brown	A-Wall
Plaid 4	Lamp_Cover
Stripes-Yellow-White	Couch

5. To see the effects of light and materials, change the visual style to Realistic.

6. Turn Sunlight on. Note the new look of your file.

7. Save and close.

11.5 CREATING NEW MATERIAL LIBRARIES AND CATEGORIES

AutoCAD's default material library called Autodesk Library is full of premade materials. To create your own material library, perform the following steps:

- Start the Material Browser.

- In the bottom left corner of the browser you can find the button illustrated below. Click it and select the **Create New Library** option:

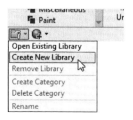

- The Create Library dialog box is displayed so you can name your new material library (*.adsklib). Type the name and click OK.

- A new, empty library is added, as shown:

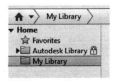

- You can fill this new library using multiple techniques, such as drag-and-drop. Go to the Autodesk Library, drag the material to the new library, and drop.

- Another way is to go to the Autodesk Library, click the material, then right-click. The following menu appears. Select the name of the new library.

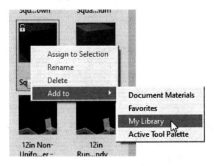

- You can organize your material library by creating categories; each category contains similar materials. To create a category, make sure you are in the right library, go to the bottom left corner of the browser, click the button shown below, and select the **Create Category** option. Type the name of the new category:

11.6.1 Reflectivity

This feature controls the reflectivity properties of the material:

The Direct option controls how much light is reflected by the material when the surface is directly facing the camera. The Oblique option controls how much light is reflected by the material when the surface is at an angle to the camera.

11.6.2 Transparency

This feature controls the transparency properties of the material:

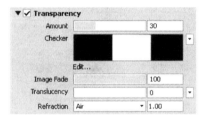

The Amount option controls the amount of light that passes through the surface. 0% is completely opaque, and 100% is completely transparent.

Image Fade, Checker, Gradient, Marble, etc., considers white to be fully transparent and black to be completely opaque.

Translucency is the transmission of light through an object or the amount of light that is scattered within the object. The value ranges from 0 (not translucent) to 100 (fully translucent).

Refraction defines the amount of distortion for an object located behind your material. Index values are: Air = 1.0, Water = 1.33, Alcohol = 1.36, Quartz = 1.46 Glass = 1.52, and Diamond = 2.3.

11.6.3 Cutouts

Cutouts control the perforation effects of the material based on a grayscale interpretation of a texture. Lighter areas of the map render as opaque, and darker areas render as transparent:

11.6.4 Self Illumination

Self-illumination means that the object emits light:

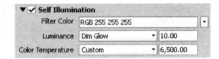

Filter Color is the color of light that is transmitted through transparent or semitransparent material, such as glass. Luminance controls the brightness of light emitted from the surface. Color Temperature controls the temperature of the color.

11.6.5 Bump

Bump simulates an irregular surface (presence of a bump or dimple):

11.6.6 Tint

Tinted material is material covered with a film or coating to reduce the transmission of light.

NOTE *You can show or hide the color of material or the texture or the objects by using the **Visualize** tab and the **Materials** panel:*

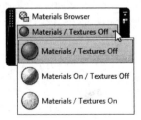

For the third method to work, the material should first be added to your drawing. If this has been done, go to the material thumbnail and right-click. The following menu is displayed. Select the **Duplicate** option:

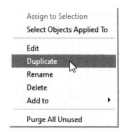

Type the name of the new material, right-click the new material thumbnail, and select **Edit** to make the necessary modifications.

PRACTICE 11-2

Creating Your Own Material

1. Start AutoCAD 2015.

2. Open practice 11-2.dwg.

3. Start the Material Browser command.

4. Create a new Generic Material, go to the Information tab, and input the following data:

   a. Name = Special Masonry

   b. Description = Material for the body of the Kitchen table

   c. Keywords = Special, Masonry, Kitchen, Table

5. Go to the Apperance tab.

6. Change the thumbnail shape to Cube.

7. Click the space next to the word Image and select the image file Brick_Uniform_Running_Orange.png. When the Texture Editor comes up, close it without changing anything.

8. Go to Bump and click on it. A dialog box is displayed. Select the image file Finishes.Flooring.Tile.Square.Circular Mosaic.png.

9. Change the amount to 1,000.

10. Assign the new material to the base of the kitchen table (make sure you are using Realistic visual style).

11. Edit the material by changing the scale of the image file Brick_Uniform_Running_ Orange.png to 120 for Width and Height.

12. Go to the Autodesk Library and select the Glass category. Drag-and-drop the **Glass Block** material onto your drawing.

13. Create a duplicate and call it Table Top.

14. Assign it to the table top. Even though it is glass, it is not transparent.

15. Edit the new material. Go to Transpareency and change it to 100.

16. Change the color to 251, 210, 75.

17. Zoom to the table top to see the material more clearly.

18. Next to the image of the transparency there is small triangle. Click it and select Waves. Close the Material Editor and Texture Editor.

19. Next to the table, create a solid box 120 x 10 x 120.

20. From the Metal category of the Autodesk Library, find Anodized – Red and copy it to your drawing. Then duplicate it and name it Decorative Wall.

21. Edit the new material.

22. Go to Cutouts, select Hexagon, and Size = 1.5, Center spacing = 1.5.

23. Assign this new material to the new solid.

24. Save and close the file.

11.7 RENDERING IN AUTOCAD

To activate the camera, lights, and material, issue the **Render** command. Rendering produces a graphical output of your model using the camera position, the effects of lights and shadows, and all assigned materials. You can save the output to a graphical file and control the graphical resolution.

There are two AutoCAD Rendering commands. They are:

- Render
- Render Region

11.7.1 Render Command

This command renders the entire scene right on your screen. It automatically opens up a new window called **Render**. The output is produced gradually by **Render Preset**, which is discussed in a few pages. The speed of the render operation depends on lots of factors, such as the complexity of the model, amount of RAM available on your machine, the processor speed, etc. You can zoom in and out using the mouse wheel. A **File** menu at the top of the screen can be used to save the rendered image to a file (*.tif, *.png, *.jpg, etc.). To issue this command, go to the **Visualize** tab, locate the **Render** panel, and select the **Render** button:

You will see something similar to the following:

11.7.2 Render Region Command

This command renders a rectangular section of the scene. This accomplishes two things:

- It takes much less time than using the Render command.

- If you changed light or a material and you want to test the effects, this command is your best choice.

No extra window pops up. Everything takes place on your current display. To issue this command, go to the **Visualize** tab, locate the **Render** panel, and select the **Render Region** button:

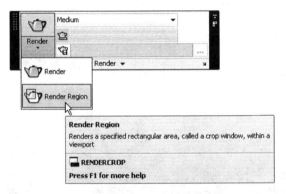

You will see something similar to the following:

11.8 RENDER PRESET

Controlling all the parameters that affect rendering with a single command is cumbersome. To make your life easier, AutoCAD offers **Render Presets** that include ready-to-use settings. They are:

- Draft
- Low
- Medium
- High
- Presentation

To select your desired Render Preset, go to the **Visualize** tab, locate the **Render** panel, and select the suitable Render Preset:

11.9 MANIPULATING OTHER RENDER SETTINGS

While in the **Render** panel, the following can also be edited:

- Render Quality
- Render Output Size
- Adjust Exposure
- Environment
- Render Window

11.9.1 Render Quality

This option adjusts the rendered quality of the model. To change the render quality, go to the **Visualize** tab, locate the **Render** panel, and select the Render Quality slider:

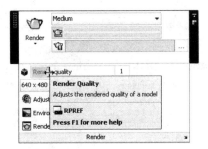

11.9.2 Render Output Size

If your image isn't clear, you may need to increase the render output size. However, as your image improves, the time needed to render and file size also increase. To change the image's resolution, go to the **Visualize** tab, locate the **Render** panel, and select the Render Output Size pop-up list:

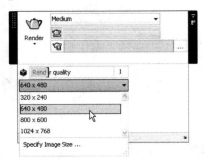

11.9.3 Adjust Exposure

This command makes adjustments to the rendered view without reissuing the Render command. This command changes any of the following settings:

- Brightness
- Contrast
- Mid tones
- Exterior Daylight
- Process Background

To issue this command, go to the **Visualize** tab, locate the **Render** panel, and select the **Adjust Exposure** button:

The 10 existing pre-made visual styles don't leave much necessity for the creation of new styles, but the options available are presented here so you know which parameters can be controlled. To create a new visual style, use the same list but select the last option, **Visual Styles Manager**:

The Visual Style Manager palette is displayed:

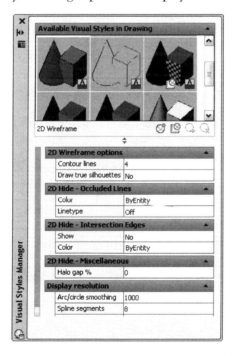

Click the **Create New Visual Style** button to create a new visual style:

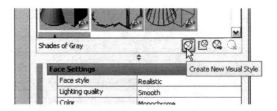

The following dialog box is displayed:

Type the name and description for the new visual style and click the **OK** button. A new visual style is added and is ready for manipulation. You can change the following:

11.10.1 Face Settings

Below are the face settings:

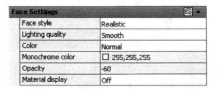

Change the **Face Style** by selecting one of the following:

- Realistic: same as the realistic visual style
- Gooch: a warm-cool face style similar to the conceptual visual style
- None: same as the wireframe and hidden visual styles

Change the **Lighting quality** by selecting one of the following:

- Faceted: highlights the facets, or parts of, faces
- Smooth: the default option, shows smooth lighting on objects
- Smoothest: the best lighting setting, but effectiveness depends on other settings

Change the **Color** by selecting one of the following:

- Normal: displays faces without any color modifier

- Monochrome: displays faces as black and white

- Tint: displays faces with monochrome color with some shading

- Desaturate: same as tint, but with softer color

The Monochrome color value is grayed out. It is enabled if you change the Color setting to Monochrome.

Change the value of **Opacity** to specify whether the materials are transparent or opaque. To turn opacity on, click the following button:

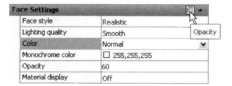

Change the **Material display** by selecting one of the following:

- Materials and Textures

- Materials

- Off

11.10.2 Lighting

Below are the lighting settings:

Change the **Highlight intensity** to control the size of the highlight on faces without materials. To control this parameter, unlock it:

Change the **Shadow display** to control how a shadow is displayed in the visual style. The available choices are:

- Mapped Object Shadows

- Ground Shadows

- Off

11.10.3 Environment Settings

This option controls whether a background is displayed or not:

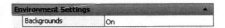

11.10.4 Edge Settings

Below are the edge settings:

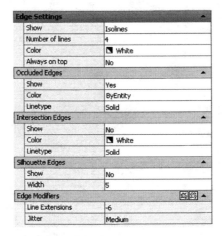

The **Show** control selects one of the following options:

- Facet Edges (controls the color)

- Isolines (controls the number of lines, color, and whether the lines are on top or not)

- None

The following picture illustrates this concept:

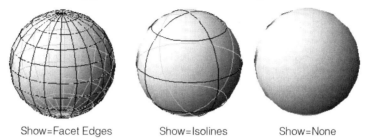

Show=Facet Edges   Show=Isolines   Show=None

Occluded Edges (works only if Edge = Facet or Isolines) controls if obscured edges are shown or not, their color, and their linetype. The following picture illustrates this concept:

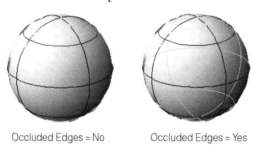

Occluded Edges = No   Occluded Edges = Yes

Intersection Edges (works only if Edge = Facet or Isolines) specifies the color and linetype of the intersection line of multiple objects. The following picture illustrates this concept:

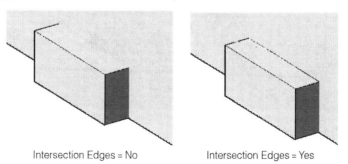

Intersection Edges = No   Intersection Edges = Yes

Silhouette Edges specifies the width of the outer edges of an object, and whether they are shown or hidden. An acceptable silhouette value is 1-25. The following picture illustrates this concept:

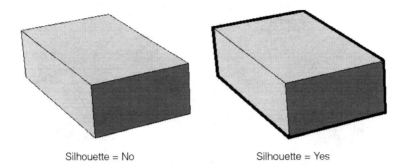

Silhouette = No   Silhouette = Yes

Edge Modifiers mimic hand-drawn lines. To control these parameters, unlock them. See the following illustrations:

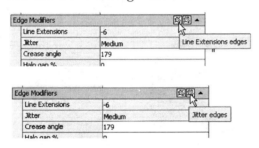

Input values for extension and jitter (High, Medium, Low, or Off). The following picture illustrates this concept:

Line Extensions = 6
Jitter = Low

After you have edited the parameters, you can ask AutoCAD to apply these settings to the current viewport:

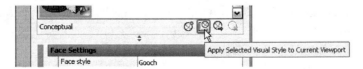

PRACTICE 11-4

Creating Visual Styles

1. Start AutoCAD 2015.

2. Open Practice 11-4.dwg.

3. Create a new visual style and call it "Special."

4. Change the following parameters:

   a. Under Face Settings:

- Face Style = Gooch
- Lighting Quality = Faceted
- Color = Monochrome
- Opacity = 70

   b. Under Edge Settings:

- Show = Facet Edges
- Occluded = Yes
- Intersection Edges = Yes
- Silhouette Edges = Yes
- Silhouette Edges Width = 5
- Line Extensions = 5
- Jitter = Low

5. Apply these settings to the current viewport.

6. Note the difference between the new style and Conceptual.

7. Save and close the file.

11.11 ANIMATION

Animation in AutoCAD is the creation of a movie of a walk-through or fly-by of a model. You can do this using the Animation Motion Path command.

First, draw a camera path (using line, polyline, spline, etc.) and a target path. You may use points instead, but the output is more compelling when paths are used.

To issue this command, go to the **Visualize** tab, locate the **Animations** panel (this panel may be hidden by default), and select the **Animation Motion Path** button:

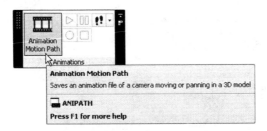

The following dialog box is displayed:

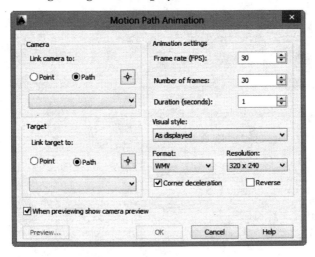

Under **Camera**, you can attach a camera to a **Point** or a **Path**. Use the small button on the right to select one. Under **Target**, do the same. This step is optional. Under **Animation settings**, change any of the following:

- **Frame rate (FPS—Frames Per Second)**. Default value is 30 FPS.

- **Number of frames**

- **Duration** (in seconds)

These three values are interdependent. Knowing two automatically determines the third. Specify the Visual Style to be used in the animation. (The list contains rendering options. If a rendering option is selected, the time required to produce the movie is quite lengthy.)

Specify the file **Format** of the output file. You can choose one of four file formats:

Specify the Resolution of the output file (higher resolution results in longer movie production time):

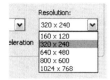

Specify if the speed of movement around corners should be reduced (Corner deceleration). Finally, select if you want to create the movie in reverse order, starting from the end of the path.

Click **Preview** to see a preview of the movie before the real processing work begins.

To wrap it up, click **OK**. AutoCAD asks you to specify the name and location of the movie file, and then the real processing work begins. It can take anywhere from a few minutes to a few hours, depending on your choices.

PRACTICE 11-5

ON THE CD

Creating Animation

1. Start AutoCAD 2015.

2. Open practice 11-5.dwg.

3. Thaw the Target Path layer.

4. Start Animation Motion Path, and specify the camera point as shown:

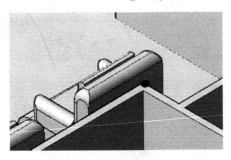

5. Set the target path to the polyline thawed.

6. Set the total to 10 seconds.

7. Set visual style to Realistic.

8. Set the file format to AVI.

9. Set Resolution to 640 x 480.

10. Name the file "Fixed Camera."

11. Freeze the Target Path layer and thaw the Camera Path layer.

12. Link the camera to the path and link the target to None.

13. Set the duration of the video to 20 seconds.

14. Set the file format to MPG, leaving all other settings as is.

15. Name the file "Moving Camera."

16. Compare the AVI file size and MPG file size. What do you think?

17. Save and close the file.

NOTES

CHAPTER REVIEW

1. There are three methods to assign materials to objects. One of them is "Attach By Layer."

   a. True

   b. False

2. Which one of the following is not related to visual style settings?

   a. Face color

   b. Intersection Edge color

   c. Highlight size

   d. Face style

3. Which one of the following is not related to Render in AutoCAD?

   a. There are two types of rendering in AutoCAD.

   b. You can control render output size.

   c. You can control render output file size.

   d. Render using different presets.

4. MPG, AVI, and ASF are among the movie file formats that AutoCAD supports.

   a. True

   b. False

5. _____ allows you to render a rectangular section of the scene.

6. _____ is the visual style that mimics a hand drawing.

CHAPTER REVIEW ANSWERS

1. a

3. c

5. Render Region

Index